NCRP REPORT No. 150

Extrapolation of Radiation-Induced Cancer Risks from Nonhuman Experimental Systems to Humans

Recommendations of the
NATIONAL COUNCIL ON RADIATION
PROTECTION AND MEASUREMENTS

Issued November 18, 2005

National Council on Radiation Protection and Measurements
7910 Woodmont Avenue, Suite 400/Bethesda, MD 20814-3095

LEGAL NOTICE

This Report was prepared by the National Council on Radiation Protection and Measurements (NCRP). The Council strives to provide accurate, complete and useful information in its documents. However, neither NCRP, the members of NCRP, other persons contributing to or assisting in the preparation of this Report, nor any person acting on the behalf of any of these parties: (a) makes any warranty or representation, express or implied, with respect to the accuracy, completeness or usefulness of the information contained in this Report, or that the use of any information, method or process disclosed in this Report may not infringe on privately owned rights; or (b) assumes any liability with respect to the use of, or for damages resulting from the use of any information, method or process disclosed in this Report, *under the Civil Rights Act of 1964, Section 701 et seq. as amended 42 U.S.C. Section 2000e et seq. (Title VII) or any other statutory or common law theory governing liability.*

Disclaimer

Any mention of commercial products within NCRP publications is for information only; it does not imply recommendation or endorsement by NCRP.

Library of Congress Cataloging-in-Publication Data

Extrapolation of radiation-induced cancer risks from nonhuman experiment systems to humans.
    p. cm. — (NCRP report ; no. 150)
"Issued November 2005,"
Includes bibliographical references and index.
ISBN 0-929600-86-X
    1. Radiation carcinogenesis. 2. Radiation—Toxicology. 3. Animal models in research. I. National Council on Radiation Protection and Measurements. II. Series.
    RC268.55.E97 2005
    616.99'4071—dc22
                                          2005031014

Copyright © National Council on Radiation
Protection and Measurements 2005

All rights reserved. This publication is protected by copyright. No part of this publication may be reproduced in any form or by any means, including photocopying, or utilized by any information storage and retrieval system without written permission from the copyright owner, except for brief quotation in critical articles or reviews.

[For detailed information on the availability of NCRP publications see page 251.]

Preface

This Report reviews the scientific issues associated with the extrapolation of radiation-induced cancer risks from nonhuman experimental systems to humans. The basic principles of radiation effects at the molecular and cellular level are examined with emphasis on comparisons among various species including humans. These comparisons among species are then continued for cancers of similar cell types in the same organ system. Risk estimates are made from an observed level of effect as a function of organ dose. The major organ systems are individually considered. Extrapolation models are reviewed and include external and internal radiation exposures.

At the beginning of the nuclear age there was no idea what risks workers would face in the handling of such substances as plutonium. The only option was to rely on experimental animal data and extrapolation. This effort, together with good health physics practices and medical surveillance resulted in, by and large, a well-protected workforce.

Many experimental animal studies were undertaken shortly after World War II. One aim was to understand the biological effects of radiation, and, hopefully, the mechanisms of the effects. Another aim was to determine the influence of such factors as dose rate, radiation quality, gender, and age at exposure.

Much has been learned and the general information has been incorporated into efforts associated with risk estimations. More specifically, the experimental data have been the basis for selecting dose and dose-rate effectiveness factors (DDREF) and radiation quality factors. These factors are used to moderate risk estimates either for patterns of irradiation or types of radiation for which there are inadequate data.

The risk estimates used in radiation protection throughout the world come almost entirely from the atomic-bomb survivors, who were acutely exposed to high-dose-rate gamma rays, not at all like the industrial (*e.g.*, uranium miners') exposures. Hence, the concern for the influence of dose rate and fractionation. An additional concern is the lack of appropriate data for estimating risk to people exposed in space missions. Again, we must rely on experimental studies.

Much remains to be done. It is believed that if more relevant data could be obtained to develop acceptable methods of extrapolation across species, risk estimates could be improved. With an understanding of what information is necessary to undertake extrapolation, it would be possible to make better use of the considerable body of data on cancer induction by radiation.

This Report includes a discussion of nontargeted radiation effects that potentially influence dose-response characteristics of cells and tissues at low absorbed doses. These nontargeted effects include bystander effects, genomic instability, and adaptive radiation responses, all of which are subjects of presently active research. It is anticipated that future NCRP reports will analyze the influence of these factors on radiation dose-response characteristics and variations in radiation response(s) among species.

This Report gives an account of the steps by which a group of researchers has advanced the pragmatic and theoretical approaches to extrapolation of estimates of risk from radionuclides and external radiation. The Report identifies the problems in extrapolating the current data from, for example, mice to humans. It provides examples of using Bayesian statistics to successfully estimate DDREF values for humans from data for mice, and also provides a measure of uncertainty for this estimate. The Report also shows how a defensible quantitative estimate of radiation injury can be determined for humans even when the exposure data of interest are either lacking or of poor quality, and how mortality data for laboratory animals can be used to predict age-specific radiation-induced risks for humans for general endpoints like life shortening, all cancers, and selected subsets of cancers involving homologous tissues.

Every example of an interspecies prediction of radiation-induced mortality contained within this Report was made in the absence of an extensive understanding of cellular and molecular genetic effects of radiation. These latter types of effects are important for providing a degree of confidence to use of the so-called biological models, but they were not critical, as stated above, for the empirical models, which take into account a great deal of what is presently known.

Perhaps the most encouraging aspect of the Report is the account of how the effect of life span can be extrapolated across species, and examples of doing so from mice to dogs to humans are given. The differences of life span among species has long been studied, and the possibility of using life shortening not only as an integrated index of radiation effects, but also for deriving a single value of relative biological effectiveness and DDREF for radiation protection purposes seemed worth examining. Lastly, the Report recommends research that is required to advance this important field.

This Report was prepared by Scientific Committee 1-4 on the Extrapolation of Risks from Nonhuman Experimental Systems to Man. Serving on Scientific Committee 1-4 were:

David G. Hoel, *Chairman*
Medical University of South Carolina
Charleston, South Carolina

Members

Bruce A. Carnes
University of Oklahoma
Oklahoma City, Oklahoma

Robert L. Dedrick
National Institutes of Health
Bethesda, Maryland

R.J. Michael Fry
Indianapolis, Indiana

Douglas Grahn
Madison, Indiana

William C. Griffith
University of Washington
Seattle, Washington

Peter G. Groer
University of Tennessee
Knoxville, Tennessee

R. Julian Preston
U.S. Environmental Protection Agency
Research Triangle Park, North Carolina

Consultants

Kelly H. Clifton
University of Wisconsin
Madison, Wisconsin

Scott C. Miller
University of Utah
Salt Lake City, Utah

Hildegard M. Schuller
University of Tennessee
Knoxville, Tennessee

Thomas M. Seed
Catholic University of America
Washington, D.C.

NCRP Secretariat
Morton W. Miller, *Consultant* (2004–2005)
Bruce B. Boecker, *Consultant* (2004–2005)
William M. Beckner, *Senior Staff Scientist* (1992–1997, 2000–2004)
Thomas M. Koval, *Senior Staff Scientist* (1997–2000)
Cindy L. O'Brien, *Managing Editor*
David A. Schauer, *Executive Director*

The Council wishes to express its appreciation to the Committee members for the time and effort devoted to the preparation of this Report. NCRP gratefully acknowledges the financial support provided by the U.S. Department of Energy, Office of Biological and Environmental Research.

Thomas S. Tenforde
President

Contents

Preface iii

1. **Executive Summary and Recommendations** 1
   1.1 Why Is Extrapolation Still Required? 2
   1.2 Summary of Findings 3
       1.2.1 Historical Aspects 3
       1.2.2 Neoplastic Disease 4
             1.2.2.1 Hematopoietic System 5
             1.2.2.2 Lung 5
             1.2.2.3 Breast 6
             1.2.2.4 Thyroid 6
             1.2.2.5 Skin 6
             1.2.2.6 Gastrointestinal Track 6
             1.2.2.7 Bone 7
       1.2.3 Somatic Genetic Damage at Molecular and Cellular Levels 7
       1.2.4 Extrapolation Models and Methods 8
             1.2.4.1 Toxicity of Chemotherapeutic Drugs ... 8
             1.2.4.2 Life Shortening 8
             1.2.4.3 Interspecies Prediction of Life Shortening and Cancer from External Irradiation 9
             1.2.4.4 Extrapolation of Dose-Rate Effectiveness Factors 10
             1.2.4.5 Interspecies Prediction of Injury from Internally-Deposited Radionuclides ... 10
   1.3 Conclusions 11
   1.4 Recommendations 13

2. **Introduction** 15

3. **History of Extrapolation: Nonhuman Experimental Systems to Humans** 17
   3.1 Introduction 17
   3.2 Lessons Learned from Genetic Risks 21

        3.2.1 Methods of Estimation 22
            3.2.1.1 Doubling-Dose Method. 22
            3.2.1.2 Direct Method. 23
            3.2.1.3 Gene-Number Method 23
        3.2.2 Discussion of Methods of Estimating Genetic Risk 23
        3.2.3 Role of Genetics in the Estimation of Somatic Risks 25
    3.3 Somatic Risks. 26
4. **Tissue and Organ Differences Among Species with Emphasis on the Cells of Origin of Cancers**. 38
    4.1 Introduction 38
    4.2 Hematopoietic System 40
        4.2.1 Introduction: Leukemias and Lymphomas 40
        4.2.2 Comparison of Radiation-Induced Leukemias Among Species. 42
        4.2.3 Pathology and Dose-Response Relationships . . 43
        4.2.4 Comparison of Hematopoietic Systems. 44
        4.2.5 Target Cells. 45
        4.2.6 Comparison of Cytogenetic Processes: Common or Species-Specific Patterns. 47
        4.2.7 Leukemogenesis Resulting from Gene Rearrangements 49
        4.2.8 Secondary Cytogenetic Lesions Associated with Leukemia Promotion and Progression 50
        4.2.9 Hematopoietic Cell Origins of the Putative "Critical" Genic Lesions and the Nature of Induced Genic Dysfunctions 51
        4.2.10 Cooperating Oncogenes in Lymphoid Neoplasias. 52
        4.2.11 Cooperating Oncogenes in Myeloid Neoplasias. 53
        4.2.12 Hematopoeitic Microenvironment 54
        4.2.13 Summary. 55
    4.3 Lung 56
        4.3.1 Introduction 56
        4.3.2 Adenocarcinoma 57
        4.3.3 Squamous-Cell Carcinoma. 58
        4.3.4 Small-Cell Lung Carcinoma. 59

4.3.5	Large-Cell Carcinoma 62
4.3.6	Summary 62
4.4	Breast 63
4.4.1	Histogenesis of Mammary Glands and Mammary Cancer 63
4.4.2	Hormones and Mammary Carcinogenesis 65
4.4.3	Cellular Origins of Mammary Cancer 66
4.4.4	Summary 68
4.5	Thyroid 68
4.5.1	General Background 68
4.5.2	Histogenesis of the Thyroid Gland and Thyroid Cancer 69
4.5.3	Thyroid Function and its Control 70
4.5.4	Cellular Economy of the Thyroid Gland and the Origin of Cancer ... 71
4.5.5	Summary 72
4.6	Skin 72
4.6.1	Introduction 72
4.6.2	Epidermal Cancers 73
4.6.3	Melanoma 74
4.6.4	Tumors of the Dermis 74
4.6.5	Mechanisms of Epidermal Carcinogenesis 74
4.6.6	Importance of Interactions 75
4.6.7	Summary 75
4.7	Gastrointestinal Tract 75
4.7.1	Introduction 75
4.7.2	Stomach 75
4.7.3	Small Intestine 76
4.7.4	Colorectal Tumors 76
4.7.5	Summary 77
4.8	Bone 77
4.8.1	Humans 77
4.8.2	Mice 78
4.8.3	Rats 79
4.8.4	Dogs 80
4.8.5	Summary 80

5. **Radiation Effects at the Molecular and Cellular Levels** 84
   5.1 Introduction 84
   5.2 Effects of Ionizing Radiations at the Molecular Level .. 85

       5.2.1 DNA Damage 85
       5.2.2 Repair of DNA Damage 86
             5.2.2.1 Single-Strand Breaks...............87
             5.2.2.2 Double-Strand Breaks 87
                      5.2.2.2.1 Nonhomologous End-Joining 88
                      5.2.2.2.2 Recombination Repair......90
             5.2.2.3 Base Damage Repair 91
       5.2.3 Characterization of Genes (Enzymes) Involved in DNA Repair....................93
       5.2.4 DNA Repair and Cell-Cycle Progression...... 95
       5.2.5 Genetic Susceptibility to Ionizing Radiations...98
       5.2.6 Conclusions.............................100
  5.3 Effects of Ionizing Radiations at the Cellular Level ..101
       5.3.1 Point (or Gene) Mutations 101
       5.3.2 Chromosome Aberrations and Deletion Mutations 102
       5.3.3 Use of Mechanistic Data on Mutation and Chromosome Aberration Induction 107
       5.3.4 Cell Killing 108
       5.3.5 Potential Confounders of Dose-Response Curves 109
             5.3.5.1 Bystander Effects.................109
             5.3.5.2 Genomic Instability 110
             5.3.5.3 Adaptive Responses 110
       5.3.6 Genetic Alterations in Tumors in Humans and Rodents 111
             5.3.6.1 Oncogene Activation...............111
             5.3.6.2 Tumor-Suppressor Genes...........117
       5.3.7 Conclusions.............................120

6. **Extrapolation Models** 122
  6.1 Interspecies Correlations of Chemical Toxicities 122
       6.1.1 Introduction 122
       6.1.2 Acute Toxicity..........................122
       6.1.3 Chronic Toxicity 125
  6.2 Interspecies Prediction of Summary Measures of Mortality: Relative Risk Models 127
  6.3 Interspecies Correlations of Radiation Effects 130
       6.3.1 Introduction 130
       6.3.2 Predictions of Radiation-Induced Mortality ...130

- **6.3.3** Example of Interspecies Prediction for Single Exposure 135
- **6.3.4** Conclusion 138
- **6.4** Interspecies Prediction of Age-Specific Mortality 138
  - **6.4.1** Introduction 138
  - **6.4.2** Background and Justification for Interspecies Predictions 139
  - **6.4.3** Continuous Exposure: Mice to Dogs 140
  - **6.4.4** Single Exposure: Mice to Dogs and Humans .. 143
  - **6.4.5** Conclusion 149
- **6.5** Extrapolation of Dose-Rate Effectiveness Factors.... 149
  - **6.5.1** Requirements and Limitations 149
  - **6.5.2** Conclusion 156
- **6.6** Extrapolation of Results for Internally-Deposited Radionuclides from Laboratory Animals to Humans . 156
  - **6.6.1** Temporal Pattern of Delivery of Radiation Dose..................................... 157
  - **6.6.2** Spatial Pattern of Delivery of Dose.......... 158
  - **6.6.3** Linear-Energy Transfer (Radiation Quality) .. 159
  - **6.6.4** Internally-Deposited Radionuclides for Which Human and Laboratory Animal Data are Available............................. 160
    - **6.6.4.1** Radium-226, 228 160
    - **6.6.4.2** Radium-224 161
    - **6.6.4.3** Thorotrast® (^{232}Th) 163
    - **6.6.4.4** Radon and Radon Progeny 164
  - **6.6.5** Examples of Internally-Deposited Radionuclides for Which Laboratory Animal Data are Available and for Which Links Could be Made to Human Data 166
    - **6.6.5.1** Bone Cancer..................... 166
    - **6.6.5.2** Liver Cancer..................... 169
    - **6.6.5.3** Lung Cancer..................... 171
  - **6.6.6** Examples of Linking Risks from Laboratory Animals to Human Data 175
    - **6.6.6.1** Bone Cancer..................... 175
    - **6.6.6.2** Lung Cancer..................... 181

7. Summary 183
- **7.1** Introduction 183
- **7.2** Summary 184

7.2.1 History of Extrapolation from Nonhuman
Experimental Systems to Humans 184
7.2.2 Cells of Origin of Cancer in Different
Animal Species 185
7.2.3 Radiation Effects at the Molecular and
Cellular Levels 185
7.2.4 Extrapolation Models 186

Glossary 188

Symbols and Acronyms 192

References 194

The NCRP 242

NCRP Publications 251

Index 262

1. Executive Summary and Recommendations

The process of extrapolation, which involves projection from the known to the unknown, can be a daunting task. A quote from a leading biometrician defines what the task needs: "To be useful, extrapolation requires extensive knowledge and keen thinking" (Snedecor, 1946). In preparing this Report, the National Council on Radiation Protection and Measurements (NCRP) faced the task of evaluating extrapolation of the risks of radiation-induced cancer to humans from experimental data. There is extensive information from experimental studies at the animal, cellular, chromosomal and molecular levels that might be used in deriving approaches to the problem of extrapolating risk estimates across species. But there are also gaps in the database. Therefore, uneasiness persists in the acceptability of extrapolations from nonhuman data, with the exception of the use of data from mice in the derivation of dose and dose-rate effectiveness factors (DDREF). Furthermore, although considered a choice of necessity rather than an ideal solution, data obtained from experimental animals have been used to derive an estimate of the dose-rate effectiveness factor (DREF) and radiation weighting factors.

The overall aim of this Report is to consider the possibilities, the difficulties, and the attempts to extrapolate estimates of radiation-induced stochastic effects across species, especially laboratory animals to humans. This Report is neither a compendium of stochastic effects studies, nor a detailed account of mechanisms of the induction of stochastic effects in different species. This Report does, however, discuss some of the similarities and differences of responses to radiation at the molecular level.

This Report concentrates on life shortening and cancer, with some comments on the use of data from mice, augmented by data from humans, in the estimation of the risk of radiation-induced heritable diseases. Unless it can be shown that the aspects of the mechanisms of importance in extrapolation are similar in cancers that arise from cells that differ in type, there must be concern about pooling the data for different types of tumors in a specific organ. The fact that extrapolations of risk estimates across strains of mice are feasible (Goldman *et al.*, 1973; Grahn, 1970; Norris *et al.*, 1976; Sacher, 1966; Sacher and Grahn, 1964; Storer *et al.*, 1988) may have been because the cells of origin of the tumors were the same in the same specific organs in the different strains.

The goals of this Report were broad and included an evaluation of the full range of somatic risks, the quality and quantity of the data from which extrapolation could be projected, and existing and potential methods for the extrapolation process. A number of different methods of extrapolation of risk estimates of radiation-induced cancers had previously been proposed but there has been neither a systematic examination of the similarities and differences among species at the molecular, cellular, tissue and whole-organism levels, nor of the appropriateness of the data available to attempt extrapolations.

Data from experimental animals are used in the estimation of genetic risk and in the derivation of factors to account for the effect of dose rate and radiation quality because of the lack of human data. However, direct estimates of risk of radiation-induced cancer at specific sites in animals have not been used.

1.1 Why Is Extrapolation Still Required?

Extrapolation is still required because the available data on irradiated human populations have several important limitations. The risk estimates for radiation protection are based largely on the data from the atomic-bomb survivors, who were exposed to an acute, high dose of gamma rays, and radiotherapy patients who have been treated with high-dose-rate fractionated exposures. A large number of people have been exposed occupationally, some protracted over long periods but in complex time patterns that have made it difficult, if not impossible, to determine the effect of total doses of low-dose- rate radiation. Hence, there is the need for data from experimental studies to determine values of DREF. Most of the exposures of humans are to very small multiple exposures of high-dose-rate radiation. Diagnostic radiation is one example. In the case of occupationally exposed individuals, much of the exposure occurs at ages of reduced susceptibility, and the reduction in effectiveness is not solely due to the reduced dose rate.

There are also inadequate data of the effects of high linear-energy transfer (LET) radiations such as fission neutrons of relevant energies and heavy ions for the estimation of risk in humans. There are some experimental weighting factors, but more data are needed.

It is the aim of this Report to examine the problems and potential of extrapolation of experimental findings in laboratory animals to risk estimates in humans. It is necessary to determine the criteria on which the suitability of the data for extrapolation purposes can be decided. For example, consideration is given to whether risk estimates of cancer induction should be based on cancers originating in the same types of cells and not just the same organ.

1.2 Summary of Findings

This Report has five sections following a brief introduction. First, the history and existing methodologies of extrapolation are presented. Second, a discussion is presented of selected neoplastic disease endpoints of particular importance to humans. Third, radiation-induced damage at molecular and cellular levels is reviewed in detail as the underpinning of comparisons at the tissue and organ levels. Fourth, extrapolation models and examples are given and include a discussion of the complex fields of radionuclide toxicology and chemical toxicology is included. Fifth, is a summary of the Report's findings followed by a glossary of terms and references.

1.2.1 *Historical Aspects*

A brief review of the extrapolation of genetic risks from animals to humans revealed concerns about somatic effects, which are not readily studied by traditional genetic processes. There are substantial differences in approaches to studying genetic or somatic effects: the baselines, issues and methodologies of the two areas of investigation are different, with one important exception; it has become clear that risk analysis requires a biological commonality to link the different species. For geneticists, the commonality is simply the deoxyribonucleic acid (DNA) molecule and its associated metabolic management. For those dealing with somatic effects, there is, in addition to DNA, the common process of "dying out" (the actuarial life table), the importance of which slowly became appreciated between about 1925 and 1950. This Report recounts the attempts to extrapolate risk estimates across species, critically examines the strengths and weaknesses of these attempts, and supports and extends the contention that estimates of the effects of radiation on life or neoplastic diseases can be extrapolated across species to humans.

1.2.2 *Neoplastic Disease*

In this Report, seven tissues or organs (hematopoietic system, lung, breast, thyroid, skin, gastrointestinal track, and bone) were studied for the feasibility of extrapolation of cancer risks. These were chosen for their general importance in human cancer risk analysis and because extensive data exist from animal studies. There are distinct differences in the problems and their solutions that are encountered in undertaking risk estimation and extrapolation of risks for external and internal radiation. The studies related to external and internal radiation are reported separately. In the case of solid cancers related to external radiation, a major barrier to success for extrapolation and the testing of methods of extrapolation is the lack of data of cancer based on the specific type of cancer or the type of cell or origin. Most of the data that have been used in risk estimates, such as those from the studies of the atomic-bomb survivors used in the selection of radiation limits, are for specific organs, for example lung tumors, not squamous-cell carcinoma (SCC) or small-cell cancer.

Studies of animals, particularly dogs, have long been one of the principal sources of data for estimating the risk of the effects of exposure to internally-deposited radionuclides in humans. Such studies, for example, have been used to estimate risks of latent effects of exposures to bone, lung, liver and bone marrow. An important impetus for these studies was that there was either a complete lack or a paucity of relevant data based on human experience. The need for estimates of risk to humans resulted in the examination of approaches of how to extrapolate these results across species and the adoption of the relative toxicity ratio method (reference can be made to the glossary for a definition for this and other terms, acronyms and abbreviations). There are problems of dosimetry and complicating factors such as relocation of the radionuclide, that are specific to internal emitters; thus, internal exposures to radionuclides are discussed in Section 6.6.

The similarities of aspects of the risk of induction of cancer among species encourage the search for methods of extrapolation. Success in this endeavor will not only improve the current values of DREF and relative biological effectiveness (RBE), but also may make it possible to use the considerable body of experimental data in risk estimation. Between humans and experimental animals there are some fundamental differences that are difficult to assess, such as lifestyle, longevity and environment. Perhaps the most important problem in being able to derive and test methods of extrapolation is the fact that most epidemiological data for cancer for humans that are used in radiation risk estimation are based on cancer site and not on cell type. In the case of mice lungs, the fact that lung cancers in many strains are exclusively adenocarcinomas, and small-cell cancer is *not* found, limits the potential for extrapolation. There have been few attempts to extrapolate risks of radiation-induced solid cancers, and in the case of hematopoietic diseases, despite all the information for leukemias in humans, dogs and mice, there has been no success in doing so. This Report also considers the problem of differences in host factors among species. In breast, there are important differences in the hormonal influence on tumors among species. One of the encouraging features is the demonstration of the similarities in the genes across the species that are important in the initiation of cancers. While this is obviously important, is it sufficient to allow some method of extrapolation? Another problem is the fact that much of the suitable data for the induction of solid cancers has been obtained after exposure at one age whereas human data, such as that from the atomic-bomb survivors, are for exposures at all ages. This, of course, is more important for the tumors with a marked age dependency such as thyroid cancer. Lastly, much of the data from experimental animals is restricted to a small number of strains. This is particularly true in the case of the dog.

Some of the characteristics of the organs discussed in this Report are briefly noted in the following subsections.

1.2.2.1 *Hematopoietic System.* Leukemia has been considered to be a major oncogenic effect in irradiated human populations. Rodents, dogs and humans have nearly identical hematopoietic systems, similarities in the cell types of myelogenous and lymphocytic leukemias and reticular-cell sarcomas, and some common underlying genetic components.

1.2.2.2 *Lung.* There are significant differences between animals and humans and their susceptibilities to lung cancer induction and the predominant type of cancer (Section 4.3). Thus, simple extrapolation of radiation-induced lung cancer from animals to humans is not reasonable. Selected risk analysis may be feasible for some tumors of the same cell types when there are appropriate experimental animal models, but this is not always the case. For example, small-cell carcinomas in humans lack a counterpart in experimental animal models. The extrapolation of risks from radionuclides has been reported and is discussed in Section 6.5. Further, the role of smoking (which may interact with radiation) cannot generally be evaluated in animal models.

1.2.2.3 *Breast.* The cellular components and the major anatomic and histologic features of mammary glands are similar among humans, dogs and rodents but there are important physiologic differences, for example, in the hormonal control of growth (Section 4.4). The marked strain-dependent differences for both naturally occurring and radiation-induced mammary cancer in the mouse and rat makes these rodents very useful models for the study of molecular, cellular and tissue aspects of the mechanisms involved. There has not been a systematic and critical study of how to extrapolate the extensive data on risk estimates of mammary cancer in different strains of rats and mice.

1.2.2.4 *Thyroid.* The physiology, morphology and tumor cell of origin are comparable among different mammalian species. Extrapolation of the estimate of risk of radiation-induced cancer appears feasible but no quantitative tests have been reported. For the thyroid (Section 4.5) as for the breast, the rat is considered to be the rodent of choice for extrapolation studies.

1.2.2.5 *Skin.* Provided that the data for humans are not confounded by interactions, extrapolation is feasible, but care must be taken to restrict the effort to tumors of the same cell type. It is important to appreciate that data from humans can be confounded by interactions with other chemical and physical agents. Factors such as ultraviolet (UV) exposures and how much of the skin is exposed are important in human skin cancer induction, and thus may be difficult to address in animal models. Furthermore, nonmelanoma skin cancer such as basal-cell carcinoma (BCC) can also be caused by skin exposures to UV light; this is a typical confounding factor with the assessment of radiation risks. Melanoma is considered only briefly because of the lack of evidence that ionizing radiation is a major etiological factor. These issues are discussed in Section 4.6.

1.2.2.6 *Gastrointestinal Track.* There are many similarities in the different gastrointestinal (GI) track tumors among mammalian species. Therefore, these tumors provide an excellent resource for the study of the mechanisms of carcinogenesis. However, there are not adequate data for analysis of dose-response relationships for the induction of tumors of the same type in either humans or rodents to test the possibility of extrapolation. The induction of cancer of the GI tract requires relatively high doses in rodents (Section 4.7).

1.2.2.7 *Bone.* Most of the data for the induction of bone tumors in laboratory animals comes from studies of internally-deposited radionuclides. Permissible body burdens for a number of radionuclides in humans have been derived from studies involving dogs. Whether osteogenic tumors arise from the same cells or the same lineage in dogs and humans is not clear, but the separate data for osteosarcomas and for fibroblastic and fibrohistiocytic types of bone tumors provide an opportunity for testing methods of extrapolation. High doses of external radiation are required to induce bone tumors in humans and experimental animals, and no attempts to extrapolate the risks have been reported (Section 4.8).

1.2.3 *Somatic Genetic Damage at Molecular and Cellular Levels*

The mechanism for the induction of chromosome aberrations and mutations by ionizing radiations are currently best understood for human cells. However, similar mechanisms of induction are known to prevail across a range of species. The processes that convert radiation-induced DNA damage into genetic alterations are errors during DNA repair or replication; some damage is irreparable. The errors, as judged by radiation-induced mutation rates, are broadly similar within a factor of two across mammalian species, with much of the differences in mutation rates accounted for by differences in DNA content. Thus, on the assumption that sensitivity to mutation induction is directly reflective of sensitivity to tumor induction, an extrapolation for radiation-induced tumors that allows for this factor of two is defensible.

Multiple steps seem to lead to spontaneous and radiation-induced tumors in rodents and humans, mutations and chromosome alterations (structural or numerical) being involved at each step. But particular gene alterations involved for a specific tumor type tend to be different across species. Whether this difference is significant in terms of extrapolation is not clear. The data on radiation-induced tumors are, however, limited. Certainly a similarity would strengthen the confidence in the extrapolation. In addition, there are species-specific host factors that can alter the probabilities of tumor development from initiated cells. These factors need to be investigated further to establish how they influence radiation-induced tumor dose-response models and extrapolation across species. Additional data on the mechanism of tumor formation will improve the level of confidence in extrapolating from data on rodent tumors to human tumors.

1.2.4 *Extrapolation Models and Methods*

This Report considers separately the species extrapolation methods that have been used for external and internal exposures. For comparison purposes, a short review is first given of species extrapolation issues in chemical carcinogenesis. For external exposures both life-shortening and cancer risk estimations are reviewed including Bayesian methods for DREF estimation.

1.2.4.1 *Toxicity of Chemotherapeutic Drugs.* The correlation among mammalian species of the toxicity of chemotherapeutics provides encouragement to the radiation toxicologist for extrapolation methods. A judicious combination of pragmatism and pharmacologic chemistry has created feasible and practical approaches to the preclinical and clinical trials of anticancer drugs. The lesson here is to take advantage of the animal data collected for this purpose and look for the underlying physiological commonalities.

1.2.4.2 *Life Shortening.* Two actuarial methods of extrapolation were examined for the life-shortening endpoint. The first method (specially designed for single exposure to low-LET radiation) relies on two findings described in the Report. First, Gompertz models (*i.e.*, linear equations on a semi-logarithmic scale) used to describe age-specific death rates exhibit parallel displacements from the control that are proportional to dose (*i.e.*, can be described as a function of dose). Second, at least for the species compared in this Report (B6CF$_1$ mouse, beagle dog, and humans as represented by atomic-bomb survivors), the dose-dependent displacements of the age-specific death rates can be described by the same equation. This finding produces the desirable effect that species with good dose-response data can be used to predict dose-dependent life shortening in species for which information on radiation exposure is either poor or lacking. The second example presented in the Report describes a method that uses proportional hazard models (PHMs) to perform interspecies predictions of radiation-induced mortality. In this case, the endpoint examined is "intrinsic" mortality, which refers to causes of death that arise from within the individual. As such, intrinsic mortality and life shortening (an integrated measure of damage) are closely related. Simple PHMs were used to describe the dose response in a species chosen to be the "predictor" species. The resulting model was then used to predict cumulative survivorship [$S(t)$] curves at levels of dose observed in a "target" species, in which $S(t)$ is the value of the cumulative survivorship function at time "t," where t is the age when the deaths from whatever cause are being examined. The ages associated with the predicted $S(t)$ values were scaled by a constant (the ratio of the predictor species'/target species' median ages of intrinsic death). When confidence intervals were calculated for the empirically derived $S(t)$ curves for each dose group observed in the target species, the scaled predictions from the PHM fell within these confidence intervals. In combination, these two examples demonstrate that not only can summary measures of life shortening (*e.g.*, days lost per centigray) be predicted from one species to another, but the entire schedule of age-specific death rates for life shortening can also be successfully predicted.

1.2.4.3 *Interspecies Prediction of Life Shortening and Cancer from External Irradiation.* The methodology for interspecies prediction of cancer from external irradiation relies upon the fact that the life span of all mammalian species can be described by the same mathematical formula that describes the species life table (*i.e.*, the exponential process of dying out). Intercepts and slopes are species-specific, but the equation is the same, and all species can be "created equal" by appropriate codification of the parameters.

Two examples are presented. In one test case, data from mice exposed to protracted daily gamma irradiation are used to predict the survival of beagles subjected to comparable exposure. In the second case, data from mice exposed to single doses of gamma radiation are used to predict observed survival of the atomic-bomb exposed individuals in Hiroshima and Nagasaki. Though the life tables for the latter are not yet complete, a remarkably consistent relation is apparent for the mouse to human extrapolation. The survival of unirradiated populations of mice, dogs and humans can all be described by a single cumulative function of median age at death.

Extrapolation models involve comparisons among different species or different populations within a species. In either case, differences among populations in mortality that are not related to the cause of interest (*e.g.*, accidents, infectious disease, environmental trauma) can conceal or distort a shared species response to radiation, especially for an endpoint like life shortening, which is based on all deaths. This mortality contamination problem was solved in the above extrapolations by coupling a biologically relevant partitioning of mortality, "intrinsic" mortality in the mouse to dog extrapolation, "solid-tissue tumors" in the mouse to human extrapolation, with widely available models for survival analysis that incorporate censoring.

Host factor differences also limit extrapolation. Although mice, dogs and humans often die from identical or nearly identical causes, these deaths need to occur at identical time points within the relative life span of the species to have utility in extrapolation processes. Since background mortality risks vary by age, these species-specific (host-factor) time shifts would probably cause an extrapolation based on a relative risk model to fail.

Finally, it is necessary to account for the fact that humans are exposed to radiations at a wide range of ages, while the animal data consist overwhelmingly of exposures of young adults. Radiation-induced mortality risks are known to vary by age at exposure. This appears to raise a barrier to testing methods of extrapolation because Radiation Effects Research Foundation data on atomic-bomb survivors are largely determined by effects from doses around 1 Sv at high-dose rates. When interspecies data involve large single doses or high-dose rates, the emergence of age-related effects and species dependent pathology syndromes cause the extrapolations to fail. In summary, interspecies extrapolations of radiation-induced risk for external exposure to radiation are feasible, reasonable and reliable when performed within fairly broad levels of pathology and level of exposure. It remains to be seen if extrapolations within defined levels of pathology detail are reasonable.

1.2.4.4 *Extrapolation of Dose-Rate Effectiveness Factors.* Since human data for prolonged exposures to almost all types of ionizing radiation are insufficient for direct estimation of risks, extrapolation of dose-rate effectiveness factors (DREF) using animal data has to be considered. This Report presents one example for extrapolation with data on female BALB/c mice exposed to ^{137}Cs external gamma radiation at low- and high-dose rates. A probability density function for a DREF in mice is estimated for mammary tumors with Bayesian methods and combined with a probability density for the breast cancer risk coefficient in female atomic-bomb survivors. This results in an estimated probability density function and a risk coefficient for breast cancer in humans after prolonged exposure. This probability density also describes the remaining uncertainty about the breast cancer risk in humans after prolonged exposure.

1.2.4.5 *Interspecies Prediction of Injury from Internally-Deposited Radionuclides.* Interspecies extrapolation for radionuclide toxicity has been conducted for many years, in particular for bone cancer.

The metabolic behavior, internal distribution, and deposition of the transuranics, radium and strontium, are predictable for mice, rats, dogs and humans, and the target organ, the skeleton, also responds predictably. This has permitted the use of available data from human exposures to radium, for example, to make predictions for dogs. The approach can be extended to other nuclides for which human data do not exist but animal data do exist and for which exposure or uptake patterns are different among species.

The toxicity ratio, which is based upon the ratio of dose-response slopes for the response to two different radionuclides tested in the same species, has encouraged extrapolation of risk assessment when the internal distribution and deposition of the nuclide is similar for humans and the test species.

The availability of occupational and clinical data involving exposures to radium, radon and/or thorium has provided baseline data necessary to test the reasonableness and practicability of the approach of estimating human cancer risks based on animal cancer studies using internal emitters.

Additional methods of analysis have also been explored. A Bayesian statistical model has been used to evaluate the extrapolation of bone cancer risks to humans from ^{239}Pu. A collection of studies with rats, dogs and humans on the effects of plutonium and radium was used in this effort.

Proportional hazards modeling has been employed to analyze intraspecies studies on the induction of bone and lung tumors by several different radionuclides. The comparison of bone tumor risks from ^{226}Ra in mice, dogs and humans has also been examined by the derivation of a power function relating a time-based parameter to skeletal dose rate.

Extrapolation methodology is obviously quite different for internally-deposited radionuclides than for external radiation exposures. For both situations, the availability of some reliable human data is critical to the development of reasonable extrapolation procedures. The human database provides (1) an essential feedback to plan laboratory studies and (2) a target against which new methods can be evaluated.

1.3 Conclusions

The extrapolation of some radiation-induced risks from laboratory animals to humans is feasible. Extrapolations have been performed for many years for external radiation exposures and internally-deposited radionuclides, though different methods must be used for the different patterns of radiation exposure.

The success of interspecies extrapolation rests upon the most basic common factors, which entail somatic-genetic aspects of damage and repair of the DNA molecule. This baseline then supports the consistent pattern of mortality seen among mammalian species. The mortality pattern then provides a quantitative, analytical basis to compare and unify the responses of different species to radiation injury.

The analyses in this Report on how data for life shortening can be used to extrapolate risks across species are very encouraging. It has been shown how the ratio of the median survival in control populations, based on a Gompertz distribution, can be used to adjust for the differences in life span among mammalian species. When these adjustments are used, the age patterns of radiation-induced mortality are markedly similar, thus allowing the extrapolation of risk to be made. There are, however, two differences that have to be considered. First, the fact that the animal data have been derived from animals that were exposed at the same age, whereas, the data for humans are from a general population in which persons of all ages were exposed. Second, the animal data are from populations with a restricted gene pool and with susceptibility to certain tumors, whereas, human populations are considered to be heterogeneous. There are suggestions as to how to overcome these differences. Even with these remaining issues it is suggested that it would be better to use life-shortening data for selecting values for DREF and RBEs because they are more representative of the total radiation effect than the data currently used that are restricted to a small number of relevant tumors.

It has been demonstrated that, at least for external exposures, a life-table-based model will provide an accurate interspecies prediction of death rates when summed across all causes of mortality or from all solid-tissue tumors. However, extrapolation of specific tumor mortality still requires some development. The findings for all-cause mortality support the recommendation that animal data on the effects of high-LET radiation (*e.g.*, neutrons) can be reasonably used to predict total mortality in human populations where few analytically useful data for humans exist.

The approach of extrapolation based on life shortening, the relative toxicity ratio and one of the proposed approaches to extrapolation of risk of solid cancers induced by external radiation depend on relative risk.

The extrapolation of responses to internally-deposited radionuclides has made extensive and practical use of a broad base of human experience. The experience of evaluating response and metabolic parameters in animals has led to sensible extensions to other radionuclides in test species with extrapolation back to humans. This type of extrapolation modeling seems effective in those cases where uptake, distribution, deposition, and pathologic response need to be accommodated by the prediction model.

1.4 Recommendations

1. There has been productive work on the comparison of radiation-induced life shortening among animal species and its extrapolation to humans. The current values of DDREF and radiation weighting factors are based on experimental data but the animal tumor data are considered inadequate; it is recommended that the use of the life-shortening data be considered. The studies of life shortening suggest that the data for this effect might be used with advantage to derive values for DREF and RBEs because they provide an integrated index of both cancer and noncancer effects. Such data should be obtained from appropriate regimens of protracted but terminated exposures. It is also important to focus on specific cancer types. In these analyses, the appropriate cell types and cellular events should be linked with the animal pathologic response. By incorporating molecular information, biodosimetry and radiation effects at the molecular level, uncertainties should be reduced and the quality of animal extrapolations enhanced.
2. It is important for risk extrapolation that the results from animal studies are archived, maintained and made available to researchers. This would involve good documentation in electronic form for the individual studies and identification of possible problems with the studies. This resource then could be used by more investigators concerned with the issue of animal extrapolation.
3. Develop both human epidemiological and experimental animal data that are based on specific types of cancer and not just cancer site.
4. Continue to develop and compile information about the comparative aspects of the mechanisms of radiation-induced cancer.
5. Develop approaches to extrapolation that take into account the problem of age-dependent susceptibility and differences in the heterogeneity of human and experimental animal populations.

6. Test the hypothesis that susceptibility for induction of cancer by radiation is related to the background rate.
7. Extrapolation of risk estimates from experimental systems are still required for radiations such as mid-energy neutrons and heavy ions and this entails obtaining relevant animal data. Furthermore, if values of DDREF and radiation weighting factors continue to be based on experimental data, additional data for life shortening and cancer should be sought.
8. To conduct space flights safely, it is important to obtain information about the potential adverse health effects from heavy-ion exposures. There are limited animal studies in this area, and it is recommended that attempt to determine the predictive value of neutron exposures for estimating the risk of heavy ions be continued. It would be important to validate these predictions by conducting whole animal studies using actual heavy-ion exposures.
9. The use of high-dose diagnostic radiology and nuclear medicine with both acute and chronic exposures places large numbers of patients at potential risk. It is recommended that animal and human data be used to estimate future risks of these procedures.

2. Introduction

Epidemiological studies have provided the basis for quantitative estimations of radiation-induced cancers in humans. The individuals exposed in these studies received mainly acute or fractionated low-LET exposures of gamma or x rays. For high-LET radiation, radon exposures have been studied in mining populations as well as in homes. Also the effects of radium and thorium in patients and occupationally-exposed populations have been assessed.

From these data, dose-response estimates have been developed. However, there is a continuing need to understand and estimate the effects of continuous exposures and other dose-rate patterns as well as the effects of particular radionuclides and ion particles. For example, proposed space travel requires estimation of health effects from heavy-ion exposures. Epidemiological studies do not have sufficient numbers of exposed humans to provide risk estimates for these types of exposures. What has been done in the past is to use experimental animal systems to provide the needed risk estimations. This has been particularly true for DREF, radiation weighting factors and exposures to materials such as plutonium.

Previous NCRP reports either used the results of animal extrapolations or directly carried out such extrapolations. NCRP Report No. 128, *Radionuclide Exposure of the Embryo/Fetus* (NCRP, 1998), made use of the animal results but expressed concern with the validity of the extrapolations, while Report No. 64, *Influence of Dose and Its Distribution in Time on Dose-Response Relationships for Low-LET Radiations* (NCRP, 1980), directly made calculations based on animal data.

This Report reviews the usefulness of experimental animal data in the prediction of human oncogenesis from exposures to ionizing radiation. Several sections address the relevant issues.

Section 3 provides a review of the statistical approaches used in the extrapolation of data from nonhuman experimental systems to humans.

Section 4 presents a consideration of tissue and organ differences among species with an emphasis on those cells that are the origin of the radiation-induced cancer. The section is further divided by organ systems, with discussion of the significance of each particular organ in terms of radiation carcinogenesis and its relation to extrapolation to human risk.

Section 5 is concerned with radiation effects at the molecular and cellular level. Specific types of DNA damage and repair as well as genetic susceptibility are reviewed. Mutations and chromosomal aberrations as well as specific tumor genes are considered with respect to similarities in mammalian species. This Section also considers the effect on the development of cancer by changes in gene expression and the extracellular matrix.

Section 6 deals with specific quantitative extrapolation models. It begins with a discussion of extrapolation for chemically-induced cancers. The section describes methods used for prediction among species of the carcinogenic effects of external radiation. This includes the use of Bayesian methods for estimation of dose-rate effects. The section concludes with a discussion of extrapolation of the effects of internally-deposited radionuclides in laboratory animals to humans.

3. History of Extrapolation: Nonhuman Experimental Systems to Humans

3.1 Introduction

The study of the mechanisms of biological effects of radiation have involved research at the whole-animal, tissue, cellular, chromosomal and, lately, molecular levels. These studies have elucidated many aspects of the effects of radiation but many important questions remain, especially some that are important in the determination of risk estimates required for radiation protection. The estimate of risk of radiation induction of germ-cell mutations that result in inherited diseases, and of radiation induction of cancer are fundamental for setting appropriate radiation limits for the general and working populations. Risk estimates of cancer induction by acute low-LET radiation based on the study of humans have become increasingly more accurate. However, the risk posed by exposure to low-dose-rate, fractionated, or protracted irradiation cannot be accurately determined from available data from studies on humans, and estimates must rely on experimental systems. Also, estimates of the carcinogenic or genetic effects of radiations such as neutrons and heavy charged particles, cannot currently be based on human experience. Despite studies on large populations, such as the atomic-bomb survivors, estimates of genetic risks depends in large part on studies of the mouse.

Extrapolation of risk estimates across species has been used in various aspects of protection against the effects of chemical agents as well as radiation. In the case of chemical and chemotherapeutic agents, the determination of the toxicity and threshold levels have depended on animal studies. Schneiderman *et al.* (1975) wrote a guide to a road map of how one might get "from mouse to man." Despite the "detours, chuckholes, swamps, quagmires and dead ends that may be encountered," they concluded that it was possible to extrapolate from mouse to humans, not with precision but at least usefully. Data from animal experiments remain the basis of the estimates of setting toxicity and threshold levels, and these approaches are discussed in this Report.

The need for data on the effect of radionuclides became clear early in the days of development, testing, deployment and use of atomic weapons and nuclear power. How were safety standards for workers in the weapons and energy industries to be set without adequate data from human experience?

There was some knowledge of the effects of exposure to radium dating back to the Curies and the technical and medical staffs involved in radium therapy. An increasing amount of data, especially, on the induction of bone tumors, was revealed from study of radium dial painters that spanned many decades (Evans, 1969; Fry, 1998; Martland, 1929; Rowland, 1994; Stannard, 1988). Studies of the exposure to radon in uranium miners has provided estimates of the induction lung cancer (IARC, 2001; NAS, 1999).

The U.S. Atomic Energy Commission set about solving the need for data to assist in setting protection standards for nuclear workers. It was in this endeavor that studies using dogs were added to those on mice. Claus (1976), the Special Assistant to Shields Warren, Director of the Division of Biology and Medicine, recorded in a flamboyant style the events that led up to the beagle project at the Radiobiology Laboratory at the University of Utah to study radionuclides. "... the Great White Father... and his name was Warren, and the Father turned to his noblest son John Bowers" (Director of the Radiobiology Laboratory in the early 1950s). And he said, referring to plutonium and radium, "go thou hence John and bring light unto the benighted. And John did and he called upon his henchmen Evans, Brues, Eisenbud, Brandt and Claus and together with their Great White Father they gathered ... and enunciated the Dogma, that as radium in man shall be unto radium in the dog, so shall plutonium in man be as plutonium in the dog and the name of the dog will be the beagle. Now these mighty words were hailed abroad as one giant step for mankind—but they were a very small step for the beagle." This is the concept that is known as the relative toxicity ratio and that was the first approach to quantitative extrapolation across species of risk estimates for radionuclides.

Needless, to say, this attractive but somewhat fanciful historical account left out some salient steps. The first studies on the comparative carcinogenicity of ^{239}Pu and ^{226}Ra were carried out by Brues (1951) and Licso and Finkel (1946) at the Argonne National Laboratory (ANL) using mice, rats and rabbits (see also Mays *et al.*, 1986a). In 1943, Evans introduced the concept of dosimetry ratios that subsequently became termed as "toxicity ratios." The toxicity ratio is analogous to the RBE introduced by Failla and Henshaw (1931). In 1950, the toxicity ratio of ^{239}Pu/^{224}Ra in rodents was used to derive the maximum permissible body burden of 0.04 C of ^{239}Pu for occupational exposures (Brues, 1951; Langham and Healy, 1973). It might be noted that the follow-up studies of the workers suggest that the limits set were wise ones. The experimental studies on radionuclides are discussed in this Report.

In 1980, NCRP Report No. 64 (NCRP, 1980) examined all the available data on the effect of dose rate on a wide range of biological systems, including cancer and life shortening in experimental animals. Values for the DREF were determined as the ratio of the effect per unit dose of radiation at a high-dose rate and the effect at a low-dose rate based on linear regression coefficients from dose-response curves for a small number of types of tumors in three strains of mice. Other data from rats and dogs were examined. It was concluded that the values for DREF were between 2 and 10 depending on the endpoint. In 1991, the International Commission on Radiological Protection (ICRP) included low dose and designated the factor as low DDREF, which appears to imply a curvilinear response. ICRP chose a value of two for DDREF (ICRP, 1991). If the responses on which the DDREF is based are curvilinear, what is the reason for the apparent linearity of total cancers as a function of dose for the data from the atomic-bomb survivors (Preston *et al.*, 2003)? It should be noted that the value of DDREF was considered a major source of uncertainty in the estimate of risk for radiation protection purposes (NCRP, 1997).

These examples of the use of experimental data involve extrapolation of risk estimates across species. In the radiation-risk estimates used in radiation protection the estimates are transferred from the Japanese population (a form of extrapolation) to derive international dose limits. This has been performed using two models, the additive and multiplicative (UNSCEAR, 1994). The additive model assumes that the average excess risk, in any population, given the same distribution by dose, gender and age at exposure will be the same for similar follow-up periods. The multiplicative model assumes that the ratio of excess risk to baseline risk at any age is invariant over populations with different baseline risks. The baseline risks for various cancers varies among populations.

Studies of extrapolation across species are needed because there are insufficient data for humans to estimate: (1) hereditary effects; (2) the influence of dose rate, fractionation and protraction; (3) the effect of high-LET radiations, with the exception of alpha particles; and (4) how to transport risk estimates across populations as well as species.

The influence of ionizing radiation on life shortening and the comparative effects among strains of mice and between mice and dogs have provided insights into problems involved in extrapolation of estimates of radiation risks across species. Studies of life shortening began early in the atomic age (Brues and Sacher, 1952; Sacher, 1950a; 1950b). At a symposium on the delayed effects of whole-body radiation in 1959, George Sacher suggested that the problem of long-term effects from animals to humans could be separated into two subsidiary problems. The first is the determination of the mechanism of the radiation response and the second is to find ways to estimate the relevant species parameters (Sacher, 1960).

In the early days of studies of radiation-induced life shortening it was considered that the effect was nonspecific but experimental evidence indicated that the effect was due to increase in the same types of cancer and normal tissues that caused mortality in unexposed populations. Exposure to radiation increased the probability of an excess of cancer and did so in a comparable manner in all species studied. The early studies identified a relatively invariant life shortening of ~28 d per 1 Gy in six genetically different strains of mice (Grahn, 1959). If the dose rate of the protracted exposures was less than ~0.10 Gy d^{-1} the life shortening was decreased to ~4 d per 1 Gy (Grahn and Sacher, 1968). More recent studies have provided a comparison of life shortening in mice, dogs and humans which will provide a guide to extrapolation of risks for cancer at specific sites (Carnes *et al.*, 2003).

The need to establish exposure standards for humans under relevant exposure conditions is the principal motivation behind interspecies modeling in the risk analysis framework. At present, available data may be insufficient for some exposure conditions (*e.g.*, protracted occupational exposures) and nonexistent for others. For example, humans have been exposed to high-LET radiations (*e.g.*, neutrons and heavy ions); however, health effects have been assessed except in a narrow set of circumstances. Therefore, risk estimates of cancer incidence, prevalence and fatality cannot be made based on human exposures. For similar reasons, it is impossible to estimate directly the influence of changes in dose rate and dose protraction on the induction of cancer in humans by radiation. It is probable that the determination of the influence of these factors and estimates of the effects of high-LET radiations will continue to be derived mainly from the results on experimental animals. It is essential, therefore, to establish how the estimates of risk based on data from experimental animals, may be extrapolated to humans.

The extrapolation of biological effects across species can be addressed at empirical, mechanistic and theoretical levels. The purpose of this Section is to provide a historical context for the quantitative methods and the biological endpoints used for interspecies comparisons. In Section 5, a biological rationale for interspecies extrapolation at the cellular and tissue levels of organization is examined. In the final section of the Report, the emphasis will shift to discussions and demonstrations of methods that have been used to estimate radiation-induced risks for humans derived from data for laboratory animals.

The history, the progress, and the current success in extrapolation of risks across mice, dogs and humans are described in this Report. It is hoped that this Report will stimulate thought and research that will result in acceptable methods of extrapolation across species. It is certain that such studies will help our understanding of what underlies the probability of the induction of cancer and noncancer effects by radiation.

3.2 Lessons Learned from Genetic Risks

This Report will not address the issue of interspecies extrapolation of genetic risks, which is a topic beyond the scope of this Report. It is reasonable, however, to ask if anything might be learned (about interspecies extrapolation of somatic risk) by examining the procedures and problems geneticists faced, since they have had to deal almost exclusively with data from laboratory animals and with theoretical concepts.

Concern about the possible genetic consequences in humans exposed to ionizing radiation date to the late 1920s, when the correlation between radiation exposure and mutation frequency was clearly established (Muller, 1927). This concern became a major consideration during World War II, and several large-scale laboratory studies were initiated under the auspices of the Manhattan Project (Charles, 1950; Deringer *et al.*, 1954; Spencer and Stern, 1948). By the mid-1950s, concerns about somatic and germ-line radiation injuries entered public discussions of political and military policy.

The increasing frequency of the testing of nuclear weapons in the atmosphere introduced the inescapable exposure of whole nations to low levels of ionizing radiation that resulted from the fallout of radionuclides produced by the fission of uranium and plutonium. Accordingly, public concerns were addressed by the nearly simultaneous establishment in 1955 of several expert committees that were commissioned to evaluate existing knowledge and to put forth recommendations on genetic risks to governing bodies. The committees and/or reports were the Committee on the Biological Effects of Atomic Radiations, U.S. National Academy of Sciences/National Research Council (NAS/NRC, 1956), the Medical Research Council of the United Kingdom report on *The Hazards to Man of Nuclear and Allied Radiations* (MRC, 1956), and the United Nations Scientific Committee on the Effects of Atomic Radiation (UNSCEAR); the latter issued its first report in 1958 (UNSCEAR, 1958).

Forty years ago, genetic effects were the primary concern, but there were no simple methods for quantitative prediction of the nature and magnitude of genetic risks to humans following exposure to radiation. The situation was somewhat better regarding somatic effects, with an accruing body of knowledge of radiation effects from the long-time use of radiation in medical diagnosis and therapy, from industrial applications, and from the use of atomic bombs. Geneticists, however, had only the satisfaction of knowing that genetic mechanisms are essentially invariant across species.

3.2.1 *Methods of Estimation*

Over the years, several procedures have evolved as a means of estimating the genetic effects of radiation exposure in humans. These are briefly defined in the following paragraphs. A detailed history of these methods and their application can be found in UNSCEAR (1988). In this Report, the primary concern is with the methods of estimation and how these may have contributed to the development of extrapolation models for somatic risks.

3.2.1.1 *Doubling-Dose Method.* The doubling dose is that dose considered to double the average spontaneous mutation rate in the genome of a species. The reciprocal of the doubling dose is called the "relative mutation risk." When the relative risk is multiplied by the spontaneous genetic burden, the product is the expected increase in that burden induced by an absorbed dose of 1 Gy. It is assumed either that the spontaneous burden is all sustained by recurrent mutation (as for fully expressed dominant genes) or that some known portion of the burden is sustained in this way. That portion is known as the "mutation component" of the spontaneous burden and entails an additional multiplier. The doubling-dose method is often called the "indirect method."

The origin of the concept is not certain, but may have first been identified in a paper by Wright (1950), although the idea derives from a number of considerations. These include the assumption of simple proportionality between dose and response and the interest in identifying the proportion of spontaneous mutations attributable to natural background radiation. Discussions on the magnitude of the doubling dose as a result of the assessment of clustered mutations lead to the conclusion that the current estimate of genetic risk is too high (Russell and Russell, 1996; Selby, 1998a; 1998b).

3.2.1.2 *Direct Method.* Because the doubling-dose method was so indirect, it was natural for geneticists to seek a means of extrapolating risks more directly to humans from data obtained from experimental animals. The direct method employs experimentally derived mutation rates for specific classes of mutation that have clinically defined counterparts in humans and can thus be used more or less directly to estimate the risk in humans. The method was developed by Ehling (1976; 1991) and Selby and Selby (1977).

3.2.1.3 *Gene-Number Method.* This approach evolved from the concept that if the total number of mutable loci were known for the human genome along with a reasonable estimate of the average mutation rate per locus per gray, then one could predict the total number of new mutations induced by a given radiation exposure. The approach reflected Muller's concerns about the existing and potentially added load of mutations in humans and the consequences thereof in terms of the frequency of "genetic deaths" (Muller, 1950).

3.2.2 *Discussion of Methods of Estimating Genetic Risk*

Of the three methods described, the third (the gene-number method) has fallen into disuse. There were too many unknowns, and the method could not produce useful quantitative values that could be related to the known genetic burdens borne by society.

Present-day risk estimates use a mix of indirect and direct methods. There are strong proponents and rationalizations for both methods. An excellent history of their application for risk assessment is given in UNSCEAR (2001). An analysis of the strengths and weaknesses of the two methods is in Sankaranarayanan (1991a). Both methods require matters of judgment rather than measurement and thus lack estimates of variability. The direct method is not as direct as the name implies and requires complex genetic and clinical matters (NAS/NRC, 1990). The doubling-dose method suffers from inadequate knowledge of radiation-induced mutation rates for specific classes of human genetic diseases and, especially, of the magnitude of the mutation component of the vast class of genetic conditions listed as congenital abnormalities and other disorders of complex etiology (NAS/NRC, 1990). As mentioned above, there is also concern with regard to the magnitude of the doubling dose as a result of considerations of how to handle clustered mutations.

Gene expression profiling is a very new technique whereby one can determine whether there is a propensity for a patient to develop a neoplastic disease, or if the disease is already present, to gain insights into its type and state of development (West et al., 2001). The present genetic analyses indicate three possible mechanisms for the development of a disease. There can be: (1) impairment of a DNA repair pathway, (2) transformation (i.e., mutation) of a normal gene into an oncogene (i.e., a cancer gene), or (3) malformation of a tumor-suppressor gene.

This new, emerging and growing technology has involved a process known as gene profiling as undertaken in DNA microarray testing sequences. One can examine thousands of genes from a single tumor sample through analyses of messenger RNAs. An individual messenger RNA is a direct complementary representation of a portion of encoding DNA (i.e., the gene encodes a messenger, and by assaying the messenger one can decipher the gene). Such microarray analyses allow for a sampling of literally thousands of messengers, and therefore, assay for genetic bases. In this way, usually through statistical analyses, one can assay for gene expressions associated with a particular tumor type. Once tumor-specific messengers are identified, then that information can be used for screening purposes to determine if noncancer patients have the genetic potential for such a disease.

Gene expression profiling is perhaps best known for breast cancer, where two human breast cancer genes (*BRCA1* and *BRCA2*) have been identified. These genes are of the mutation type, and their individual presence indicates an increased potential for developing breast and/or ovarian cancer. *BRCA1* is a mutation on chromosome 17 and *BRCA2* is a mutation on chromosome 13.

The field of gene expression profiling is a fast moving one. The National Center for Biotechnology Information is charged with providing daily updated information in this area through its *Genes and Disease* website (NCBI, 2005). Presently, the website lists breast cancer, Burkitt lymphoma, colon cancer, leukemia (chronic myeloid), small-cell lung carcinoma (SCLC), multiple endocrine neoplasia, neurofibromatosis, pancreatic cancer, polycystic kidney disease, and prostate cancer as having a genetic basis to its etiology. There is also a wide range of similar information available from commercial sites dealing with animal diseases and microarrays by which to undertake specific analyses.

3.2.3 *Role of Genetics in the Estimation of Somatic Risks*

Somatic and genetic risk assessments progressed independently with little or no cross communication. From the mid-1950s to the mid-1980s, the data on stochastic somatic effects in humans continued to accrue and become more compelling regarding ancillary variables such as age, sex, dose and cell type for neoplastic diseases. Animal data seemed less important except for questions about the influence of dose rate and radiation quality and estimates of the risk of cancer from internal exposure to radionuclides.

Great strides were also being accomplished in genetics but, ironically, as noted by Crow (1982), ". . . the magnificent accomplishments of molecular biology have diminished rather than augmented our confidence in quantitative assessment of the human genetic risk." As Crow added, this new knowledge created new uncertainties. Yet, in radiation genetics, most of the parametric values needed for extrapolation were now available for the mouse.

Since molecular and cellular biology had not solved the problems involved in expressing human genetic risks, what was missing? At least one major item was missing, namely, reliable measures of the radiation-induced mutation rates for representative human genetic defects. While estimates of radiation-induced induction rates were developed for several indicators of genetic damage among the children of atomic-bomb survivors in Japan (Neel *et al.*, 1990), no measures were statistically significant and some were essentially zero or negative. The absence of even a single baseline mutation rate has made it extremely difficult to reduce the uncertainty of genetic risk estimates for humans.

The role of genetics then is that baseline values of induction rates derived from human populations are absolutely essential to reasonable and quantitative extrapolation modeling. Somaticists have the needed parameters, geneticists do not, at least not to the same degree of certainty as exists for induction rates for several neoplastic diseases. The present availability of transgenic mice may provide an experimental system by which radiation-induced genetic risks of multilocus chronic diseases can be quantitated for subsequent application in extrapolating to human genetic risk.

3.3 Somatic Risks

The history of species comparisons of mortality can be viewed as an effort to acquire relevant data and as an evolution in the development of endpoints used for comparison. Although the availability of data is less of an issue today than in the past, neither a consensus on the appropriate biological endpoints nor agreement on the quantitative methods for extrapolation has emerged. Pearl (1922), in his search for a fundamental law of mortality, may have been the first investigator to compare species formally. His comparisons were based on vital statistics (Pearl, 1923) arranged in standard actuarial tables.

Potentially large differences in the length of life represent the most obvious obstacle to making comparisons between species. Pearl (1922) chose to superimpose "two biologically equivalent points" within the life cycles of the organisms being compared to adjust for differences in life span. He used the elapsed time from the age where death rates reach a minimum and the age where survivorship is reduced to one individual per 1,000 to convert (normalize) time into units expressed as "centiles of the life span." Using this scaling approach, Pearl (1922) produced normalized survivorship curves for *Drosophila* and humans that were sufficiently alike to suggest that the laws of mortality were fundamentally the same for the two species. Quantitative differences between the survival curves for the two species were attributed to humans' ability to control their environment. A year later, Pearl (1923) added the rotifer (*Proales*) to his comparisons and issued a plea to the scientific community to collect mortality data on other organisms.

Using data from the field of animal husbandry and nutrition, Brody (1924) also attempted to characterize quantitatively what he called the "kinetics of senescence." Brody's model derived from Loeb and Northrop (1916; 1917) observations that a number of life phenomena, including duration of life, has a temperature coefficient, which is related to an organism's body temperature, characteristic of chemical reactions. As such, duration of life was thought to have a physiological basis determined by the time course required to complete a series of chemical reactions.

Using a variety of biological endpoints and species of animals (*e.g.*, milk production of dairy cows, egg production of domestic fowl, survival of fibroblasts in the serum of domestic fowl of different ages, wound healing in humans, Pearl's data on *Drosophila* mortality), Brody (1924) demonstrated that the behavior of these diverse endpoints through time could be described by a simple equation for exponential decline, which is termed "the law of monomolecular change in chemistry." Supplemented by examples drawn from existing mortality data for humans (*e.g.*, cancer, diseases of the arteries, and diseases of several organ systems), Brody (1924) argued that age and disease-specific death rates represented a quantitative measure of senescence and the normalized reciprocal of the death rates could be interpreted as the time course of the general "vitality" of an organism.

Greenwood (1928) attempted to provide a biological perspective on the "order of dying-out" described by a life table. Specifically, he focused on the physiological implications of the actuarial model developed by Gompertz (1825). The lack of arithmometers and the prevailing attitude of actuaries that the life table was just a working tool were cited by Greenwood as the reasons why Gompertz's generalization of his simple graduation model to be more biologically plausible (by incorporating individual heterogeneity) had been forgotten (Section 6). By example, Greenwood also demonstrated that the extension of the Gompertz equation by the famous actuary Makeham (1867), who adjusted for factors affecting mortality independent of age (*e.g.*, environmental factors), made neither biological sense nor provided any significant improvements to the graduations provided by the earlier, simpler Gompertz equation. Greenwood used the numerical uncertainties of life-table statistics, estimated in the tail of the dying-out distribution, where only few individuals remain, to question the scaling approach of Pearl (1922). Instead, he used the expectation of life from the age of minimum mortality as a scaling device for comparisons among species. Using this approach, Greenwood concluded that "we have no sound reason for thinking that the force of mortality in mice increases with age more nearly geometrically than the force of mortality in men" nor is there any "reason to think that any more complex formulation of a physiological law would describe the observed facts better than Gompertz's century-old simple formula."

Over the next decade, mortality data for a variety of organisms (*e.g.*, saturniid moth, roach, domestic fowl, mice) began to accumulate. In response to the biological and statistical criticisms by Greenwood (1928) of their original method of life-span adjustment, Pearl and Miner (1935) began expressing time as "percent deviations from the mean duration of life." This method of adjusting for life-span differences became broadly applied as ecologists began developing life tables for a rapidly growing list of organisms (Deevey, 1947).

As a larger pool of mortality data became available, Pearl and Miner (1935) came to the conclusion that there was not a "universal law of mortality" that could be applied to all species. Their effort to develop a unified paradigm for mortality failed, in part, because it became apparent that in natural populations of animals, the intrinsic causes of mortality could not be separated from the extrinsic (environmental) causes of mortality that differ widely among species. Subsequent comparative efforts shifted toward categorizing the mortality schedules of species into distinct "types" of life tables (Deevey, 1947; Pearl and Miner, 1935).

In the 1940s, attempts to characterize the biological effects of exposure to radiation began to intensify just as the search for natural laws governing duration of life had begun to wane. As might be expected, mortality data organized in life tables were used to quantify the effects of radiation exposure. Initially, attempts were directed at estimating a "tolerance" or "permissible" dose in laboratory animals, principally mice. A permissible dose was defined as a quantity of radiation that produced neither any detectable pathologic changes nor any significant shortening of the life span (Lorenz, 1950). The endpoint used for these studies was variously referred to as the mean expectation of life or the median survival time.

As the biological effects of radiation exposure became better known, a need arose to develop a link between the actuarial measures of radiation injury and the biological consequences of radiation exposure. Sacher (1950a; 1950b) postulated that "radiation initiates in organisms a lethal process that is a function of the many forms of physiological injury produced." He developed an impulse injury function, which when combined with a presumed "lethal bound" of injury, led to a metric (the integral lethality function) that according to Sacher, described a "generic mammalian radiation-injury process." The effects of radiation were assumed by Sacher to combine additively with the process of natural aging, thus "accelerating pathological tendencies but introducing no qualitatively new pathology." Under the assumption of independence, the Sacher (1950a; 1950b) model accounted for natural aging by the inclusion of a simple linear time-dependent term to the integral lethality function for radiation injury.

In a very brief passage, Sacher (1950a) introduced a quantitative relationship that would reappear in the radiation literature for decades to come and would eventually become dogma within the field of radiation biology. He began by observing that at low daily dose rates, the reciprocal difference in median survival times for a control and an exposed population was proportional to the intensity of exposure (measured in units of dose rate). With a little algebraic manipulation, the relationship of reciprocal survival times implies that the lethality function is an "always-increasing" function of time. Since natural aging was accounted for in the model and the dimensions of time entered only as a ratio, Sacher had provided researchers with an easily calculable statistic (later called the cumulant lethality function) that he argued would serve as "a purely empirical transformation in the investigation of the comparative lethality in different species, especially where these have widely different normal life expectations." In addition, the physiological concept and the subsequent mathematical formulation of the Sacher model did not depend on a specific radiation quality or pattern of exposure.

A linkage between the physiological processes and the actuarial response to radiation injury was formalized by Brues and Sacher in 1952. They envisioned injury as a process that disrupts the normal oscillations about a mean homeostatic state within an organism. They further reasoned that there must be limits (lethal boundaries) to departures from the mean homeostatic state that an organism could tolerate. Thus, the animal would die when an insult was large enough to cause a pulse in the homeostatic state to cross the lethal boundary. Brues and Sacher (1952) noted that this biological model of injury and failure leads directly to the "mathematicophysical formulation" derived by Gompertz (1825) to describe "the law of human mortality."

It is useful to note that the Gompertz distribution is but one member of a larger family known as extreme value distributions (Gumbel, 1954). Extreme value distributions have played an important role in reliability analyses within engineering as well as the biomedical sciences (Lawless, 2002). They are typically employed to describe the failure times of systems that cease to function whenever the weakest (hence extreme) component of the system fails. In the context of the Brues and Sacher model, the organism dies whenever the homeostatic control of a critical physiological process fails because of an injury process initiated by a radiation insult.

For the remainder of the 1950s, the cumulant lethality function continued to play a dominant role in the comparative analyses of radiation lethality. Through the use of median survival times, cumulant lethality functions were estimated to compare empirically the similarities and differences in the response of species to radiation injury within phases of the injury process (Sacher, 1955; 1956a). In his formal derivation of the mathematical relationship between physiological injury and mortality (the Gompertz function), Sacher (1956b) felt that his model was still "far from adequate" for "making valid inferences about effects on man in terms of laboratory experience."

As Sacher (1956b) noted, his procedure was the "first to account for mortality in terms of the statistical nature of physiological processes." However, Sacher and Trucco (1962) also noted that "we have insufficient knowledge about the nature of the fluctuation process in real physiological systems." Poorly understood dynamics of the physiologic function include such species-specific (host) factors as distance to lethal bounds and the normal behavior of oscillations around the physiologic steady state (Sacher, 1960). In addition, Sacher and Trucco (1962) noted that "the very fact of performing an observation introduces a disturbance that makes it impossible as a matter of principle to study the system's behavior with unlimited precision." As Sacher (1960) so aptly put it, "any living system, even the simplest, is a control system of a complexity and sophistication that surpasses our present ability to understand or describe."

Failla (1958) also recognized that mortality patterns conformed to the Gompertz distribution once "adulthood" was attained. The similarity of mortality patterns adjusted for extrinsic (violent) causes of death across species (*e.g.*, mouse, rat and human) was interpreted as evidence for a common aging process. Like Brody (1924) before him, Failla (1958) defined "vitality" as the reciprocal of the age-specific mortality rate. After expressing the Gompertz function in terms of vitality, he suggested that the resulting equation described the loss of vitality from a "one-hit" random process acting on the cell population of the body. The decline in the vitality curve by the end of the life span exceeded what could be attributed to a depletion in the number of cells. Failla concluded, therefore, that the vitality curve must describe a deterioration in the function of cells with age. The deterioration of function was attributed to somatic mutation, and the Gompertz aging parameter (derived from mortality data) was interpreted as an estimate of the "spontaneous somatic gene mutation rate per cell per year." With some assumptions about "generation" length and the number of genes in diploid cells, Failla's (1960) calculations suggested that the mutation rate per generation was similar across species (*e.g.*, mouse, rat, human, *Drosophila*). If true, the somatic mutation rate per unit time must be higher in short-lived animals than in animals with longer life spans. Failla (1960) concluded that "life span is determined by the inherent stability of the genetic system of a given species, which determines the spontaneous mutation rate, which in turn determines the increase in mortality rate with age (beyond middle age)."

Szilard (1959) also developed a theory on the nature of the aging process based on the concept that accumulated somatic damage interferes with the functional capability of cells. Inherited "faults" (mutations) in somatic genes whose function is critical late in the life span were viewed as the major explanation for why adults differ in length of life. While similar in concept to the Failla theory, Szilard's approach was far more extensive in transforming the theory into a quantitative form. Like Sacher's lethal bound, Szilard envisioned death occurring when the fraction of somatic cells unaffected by "hits" reached a critical threshold. Numerical relationships were developed that permitted an estimation of the surviving fraction of cells, the critical threshold, the number of somatic mutations, and the reduction in life expectancy per mutation. Szilard suggested that the magnitude of life shortening following exposure to radiation should be inversely related to the square root of the number of chromosomes for a species. As such, mouse and human should experience a similar radiation-induced life shortening when expressed as a fraction of the life span (Section 6, Figures 6.5 and 6.6).

The quantitative as well as the biological importance of the Gompertz distribution in the study of radiation effects was further enhanced by the work of Strehler and Mildvan. In a series of papers (Mildvan and Strehler, 1960; Strehler, 1959; 1960; Strehler and Mildvan, 1960), these investigators presented a Gompertz based theory of mortality and aging that, like the Sacher model, was based on disruptions of the homeostatic state of an organism. Their approach differed from that of Sacher in the functional form of the equations used to describe the disturbances of the "energetic environment" of an organism when challenged by a stress. This difference, they argued, was critical if derivative implications of the theory concerning such issues as the predicted loss of physiologic function with time and the quantitative relationship between the two parameters of the Gompertz distribution were to conform with "observation or natural law."

Even though Strehler and Mildvan never dealt directly with the interspecies issue, their stochastic model of mortality and aging deserves mention for several reasons. First, in each of their papers they provided detailed quantitative overviews of the prevailing theories of mortality and aging. Second, their theory (Strehler, 1960) addressed a major criticism concerning the relevance of the Gompertz distribution for actuarial analysis that persists to this day (Olshansky *et al.*, 1993), "namely that contrary to Gompertzian expectation, death rates level off at extreme old age." Finally, as regards radiation effects, their model (Strehler, 1959; Strehler and Mildvan, 1960) accurately predicted that the Gompertz intercept should increase in proportion to dose for single exposures, and the Gompertz slope should increase in proportion to the dose rate for continuous exposures previously noted by Brues and Sacher (1952).

Strehler (1959) also made several important observations on the biological effects of radiation compared with the effects of aging. He noted that: (1) aging effects are typically associated with post-mitotic cells while radiation primarily affects dividing cells, (2) radiation damage is primarily genetic whereas the effects of aging appear at all levels of biological organization, (3) some species (*e.g., Drosophila*) do not exhibit life shortening even after large doses of radiation, and (4) the dose required to double the mortality rate (Gompertz slope) produces a much larger increase in the mutation rate. Based on these observations, Strehler rejected the notion that radiation acts through "a general acceleration of the normal aging process."

The mid-1950s to the mid-1960s was an interesting period of scientific discovery due to the significant contributions, quantitative as well as theoretical, made by radiation biologists to the field of aging research. Sacher, coming from the field of radiation biology, and Strehler, with his gerontological background, investigated aging and the effects of radiation exposure from a physiological perspective. Derived, in part, from their research on radiation effects, the theories of Failla (1958; 1960) and Szilard (1959; with comments by J. Maynard Smith, 1959) addressed the fundamental nature of the aging process from a genetic perspective. These investigators recognized the inherent interrelatedness of the aging process and the biological effects of radiation exposure. They also exemplified the scientific benefits to be derived from interdisciplinary research.

Investigations of radiation effects continued to make extensive use of the Gompertz distribution throughout the 1960s (Berlin, 1960; Sacher, 1966; Sacher and Grahn, 1964). Linearity of the hazard function (*i.e.*, an instantaneous failure rate) on a logarithmic scale made least squares estimates of the Gompertz parameters easy to calculate. The ratio of Gompertz slopes for unexposed populations of the species being compared was proposed as a scaling constant to adjust for life-span differences (Grahn, 1970). Loglinear models relating mean after-survival (MAS) or Gompertz slopes to daily dose rate were developed (Grahn, 1970). The functional form of the dose-response models and the equation for the median of a Gompertz distribution were used to derive relationships between ratios (relative to control values) of MAS and Gompertz slopes (see Section 6.2 for details). The ratio relationships and the scaling factor for mouse to human made it possible to estimate exposure conditions that would lead to 50 % reductions in human life expectancy. Similar calculations permitted an estimation of the career exposure required to achieve the 5 y reduction in life expectancy observed for American radiologists (Grahn, 1970). The resulting estimate was in reasonable agreement with calculations based on a presumed adherence to prevailing exposure standards (Section 6.2).

A decade later, Grahn et al. (1978) developed a metric called the average lifetime mortality ratio (mortality ratio). The mortality ratio was defined as the area between the age-specific mortality curve for an exposed group and the control curve divided by the number of days between 100 d (age at first exposure) and the age at last death for the irradiated cohort. A functional relationship between the mortality ratio and life shortening expressed either in time units or as a percent reduction relative to control life expectancy was developed. Estimates of life-shortening effects derived from the animal data were transformed to human terms by a scaling constant based on the ratio of Gompertz slopes.

Loglinear models are multiplicative on an arithmetic scale. In the Grahn approach (Grahn, 1970; Grahn et al., 1978), any life-table endpoint (MAS, Gompertz slope, mortality rate) for a cohort exposed to radiation was assumed to be equal to the equivalent control endpoint multiplied by an exponential term. Dividing both sides of these equations by the control value leads naturally to a consideration of ratios. Although this approach reduces the life table down to a single summary statistic, the basic formulation is the same as that used for the relative risk model applied to the hazard function (Cox, 1972; Kalbfleisch and Prentice, 1980). In fact, Grahn et al. (1978) interpreted the mortality ratio as a measure of relative risk and noted that its invariance between species gave empirical support to the assumption that relative risk models may be more appropriate than absolute risk models for extrapolating neoplastic risks across genera and species boundaries.

Life shortening has been viewed as an important endpoint in radiation biology because it summarizes, in a single index, the cumulative effect of all injuries experienced by an organism. In a study using B6CF$_1$ mice exposed to either fission spectrum neutrons or ^{60}Co gamma rays, 85 % of the animals died with or from a neoplastic disease (Grahn et al., 1992). At the lowest exposure levels (inducing 10 % life shortening or less), essentially all of the life-shortening effect was attributed to radiation-induced cancer deaths (Grahn et al., 1972; 1978). As such, it should not be surprising that the age-specific mortality rates for "neoplasms" were also comparable for mouse and human when plotted on a scale that adjusts for differences in life span (Grahn, 1970). Data from irradiated humans in Japan and from clinical studies on the long-term effects of radiotherapy generally concur with the animal studies (UNSCEAR, 1982).

Even when "neoplasms" are parsed into specific neoplastic events, similarities between mouse and human for some life-table parameters have been observed. Storer et al. (1988) used the population structure of RFM control mice to standardize the expected frequency of tumor-caused deaths in mice exposed to ^{137}Cs gamma rays. The adjusted frequencies were then used to summarize the entire life table for a cohort by a single statistic, the weighted average hazard rate. In a modeling exercise using this endpoint, relative risk models were found to be superior to models based on the assumption of absolute risk. Based on the similarities of relative risks for tumors of the lung and the breast, Storer et al. (1988) concluded that the relative risk for tumors of homologous tissues could be extrapolated across species without adjustment for differences in life span.

At first glance, there appears to be a discrepancy between Pearl's conclusion that a fundamental law of mortality does not exist and the reasonable success of interspecies extrapolation efforts within the field of radiation biology. The paradox is resolved when the environmental conditions of the animals being compared are considered. Pearl's studies and the work of the ecologists that followed him were based on a comparison of species that experienced dramatically different levels of intercurrent mortality (exogenous mortality). The laboratory animals used in radiation studies benefited, however, from husbandry practices that included highly controlled environments where predation was eliminated and the effects of infectious diseases were minimized. These environmental conditions are far more similar to the sheltered environment and medical attention received by humans than that experienced by natural populations of animals.

General principles concerning the biological responses to radiation exposures have been used to extrapolate to humans in a qualitative sense, and have also influenced the establishment of protection standards for humans. For example, the effects of radiation are reduced when the dose rate of exposure is reduced, and increased when the LET of the radiation is increased. Similarly, an increase in dose increases the probability of effects such as cancer. However, the probability and the magnitude of effects for specific pathological events can vary between population and species. For example, uncertainties have arisen concerning the validity of transferring cancer risks from the atomic-bomb survivors in Japan to other human populations exposed to radiation because of known differences in background rates for occurrence of specific tumor types (Land and Sinclair, 1991). Finally, since the factors that determine the occurrence of cancer in the majority of tissues of any species are not fully understood, empirical methods will remain (for now) the primary tool for estimating and extrapolating risks.

The complexity of differentiating between the inherited and environmental components of cancer and quantifying the role of host factors in affecting the expression of neoplastic disease suggests that confidence in any empirical approach to extrapolation can only be gained through a greater understanding of the actual mechanisms of carcinogenesis. The fact that issues such as using the aggregate response as a surrogate for the individual effect (*i.e.*, genetic heterogeneity) and the separation of the inherited (constitutional) and environmental components of mortality have been discussed all the way back to Gompertz himself is a testimony to the scientific challenge that these issues present.

Confidence in transferring risks across species, amid the need to understand the underlying mechanisms of cancer, depends, in part, on the level of pathologic detail that is desired for extrapolation. The evolutionary conservation of genes involved in DNA repair, control of cell proliferation and differentiation, the similarity of radiation-induced lesions of DNA among various species and how such lesions are resolved or expressed, are reasons to be optimistic about the possibilities of developing acceptable methods of extrapolation of risk estimates across species.

An additional reason for believing that extrapolation of cancer risks across species may be possible is the premise that cancer and aging (senescence) share common mechanisms of molecular action and regulation (Cutler and Semsei, 1989). The logic for this premise derives from the accepted notion that cancer has a genetic basis and the evolutionary forces that affect gene expression, disease manifestation, and processes of aging are the same for all sexually reproducing species (Comfort, 1979; Kirkwood, 1992; Medawar, 1952; Williams, 1957). When survival is extended into the post-reproductive period of the life span, declines in physiologic function are well known and a proliferation of fatal degenerative diseases, such as cancer, has been predicted (Carnes and Olshansky, 1993; 1997; Olshansky et al., 1993). As Medawar (1952) so aptly stated, "innate senescence ... is in a real and important sense an artifact of domestication; that is, something revealed and made manifest only by the most unnatural experiment of prolonging an animal's life by sheltering it from the hazards of its ordinary existence." Humans, their pets, laboratory animals, and some domestic animals are the only life forms that are intentionally ushered into those ages where the diseases of senescence are expressed. Using arguments based on the ancestry of mammals and the evolutionary consequences of extended survival, Carnes et al. (1996) demonstrated that humans and laboratory animals (mice and beagles) have age patterns of endogenous mortality that are statistically indistinguishable.

However, the same genetic arguments used to predict that age patterns of endogenous mortality should be similar also imply that genetic heterogeneity would be expected to lead to species differences. In other words, no single laboratory animal should be expected to be a perfect surrogate for humans. Experience has already verified this conclusion. Extrapolation efforts have been successful for gross levels of pathology detail where the general imprint of evolutionary forces is apparent. Problems in extrapolation have occurred at finer levels of pathologic detail, presumably because of genetically determined host factor differences. Inbred strains of mice have often been used in radiation studies and differences among mouse strains in susceptibility to neoplastic and degenerative diseases are well known. However, in a collective sense, the strains and species of animals used for radiation experimentation may provide enough genetic diversity to mimic the range of response to radiation observed in humans (Section 6).

One approach to assessing the extrapolation dilemma has been to develop models that incorporate information on the mechanisms of induction and processing of radiation-induced damage, especially that at the DNA level. An understanding of mechanisms prevents potentially important factors from being ignored. The mechanisms at the molecular level of the initial events may be the same in two species, but the factors that determine the expression of those changes, and therefore the final risk, may be different.

However, it is somewhat unclear just how our current state of knowledge has enhanced our ability to make predictions for whole-animal effects from phenomena observed at finer levels of biological organization. In part this has been due to the fact that it is difficult to envision how models based solely on biological events occurring at or near the time of exposure could be expected to predict the timing and probability of disease expression occurring several decades after the initiating event(s). Thus, it is of much greater value to use the knowledge of molecular and cellular perturbations of radiation exposure to enhance the predictive value of a model. The development of biologically-based models holds promise for incorporating the pertinent molecular and cellular information, thereby reducing the reliance on purely pragmatic approaches. The challenge is to establish what data should be incorporated into biologically-based models. The current state of the field tends to emphasize the complexity of biological responses to radiation, thereby expanding the uncertainties that must be incorporated into quantitative models that presume to reflect biological reality.

The predictions and implications of models can play an important role in testing hypotheses concerning mechanisms of radiation injury and providing guidance to the design of future experiments. However, there is an inevitable trade-off inherent in the development of quantitative models derived from theory. As a model's ability to predict the behavior of a specific process within a system increases, its ability to predict the response of the system at large decreases. The rate of predictive decay is related to the complexity of the system being modeled.

While acknowledging the desirability for estimates for radiation-induced health risks based on human data, the value of the laboratory animal data for interspecies extrapolation should not be forgotten or underestimated. The relevance of the animal data for predicting human health effects associated with exposure to radiation is a well-documented scientific fact. When verification with human data is possible, risk models derived from animal data have been shown successfully to predict life-shortening, cumulative risk for a variety of neoplastic diseases, relative risks for some tumors involving homologous tissues, and even age-specific death rates for pathology endpoints commonly used in analyses of the atomic-bomb survivors. These successes give credibility and confidence to the application of animal data for predicting the effects of radiation in humans. It has been demonstrated that animal data can be used to predict age-specific mortality risks in humans, and animal data are currently used to assist in predicting RBE and DREF values and genetic risks in humans.

4. Tissue and Organ Differences Among Species with Emphasis on the Cells of Origin of Cancers

4.1 Introduction

It is thought that methods of extrapolation of radiation risk estimates across species can be tested because there are data for radiation-induced cancers in humans and in experimental animals. Such a claim is predicated on the assumption that the characteristics of a cancer of a specific tissue that are important to risk estimates are the same irrespective of the species. For example, myeloid leukemia in mice can be equated with the same type of leukemia in humans.

Estimates of the risk of radiation-induced solid cancers for radiation protection purposes are stated in terms of specific organs, such as breast, lung, etc. For risks among different species, such a comparison is probably too general. Lung tumors in mice are frequently alveologenic or are Clara cell tumors, whereas SCLC, which is not found in mice, is common in humans.

The development of a classification of human cancers has been described by Berg (1996), and there is an accepted international classification of diseases in which cancers are classified based on the type of tumor and the site (WHO, 2000). There are no comparable publications with classification numbers for tumors in experimental animals. The International Agency for Research on Cancer (IARC) has published monographs on the pathology of tumors in experimental animals [*e.g.*, *Tumors in the Mouse* (IARC, 1994)].

The classification of cancers based on the organ and not the cell of origin is unlikely to be adequate for testing methods of extrapolation of risk estimates across species. Therefore, it is necessary to base risk estimates on the cell type of the cancer in humans and the experimental animal. It is also necessary to know whether there are any differences in the factors that influence the expression of the events initiating the carcinogenic process. Such host factors as hormones, that play a critical role, may differ among relevant species.

The question of whether the mechanisms must be the same in the two species for extrapolation to be valid must be addressed. There is good evidence that because genes involved in the control of cell proliferation and differentiation, which are considered central in defining cancer, have been conserved, that the initial changes in growth control at the gene level are similar across species. Where species-specific differences in the carcinogenic process are likely to occur is in the events that are usually referred to as promotion and progression, which are stages occurring during the latent period. The latent periods of all cancers are much shorter in rodents than in humans. Clearly the rate at which the rate-limiting changes in development of cancers occurs is greater in rodents than in humans, and this is likely to be related to differences in host factors.

Differences in susceptibility, especially how humans and rodents manage DNA lesions that may lead to cancer, are discussed in Section 5. The task of extrapolation requires either that the principal aspects of carcinogenesis are similar in humans and rodents and other species, or that the differences can be accommodated by the method of extrapolation.

In this Section examples of organs for which it would be desirable to extrapolate risk estimates across species are given and a comparison is made between the types of cancer for which there are data for humans and some experimental animals. The tissues chosen to illustrate the problems involved are: breast, thyroid, skin, GI tract, lung, bone marrow, and bone.

Liver is deliberately omitted from the above listing because most of the data for humans that might be used is confounded by hepatitis B. It should be indicated, however, that NCRP Report No. 135 discusses liver cancer risk: (1) from internally-deposited radionuclides, principally clinically-administered Thorotrast® (Van Heyden Company, Dresden-Radebeul, Germany), a radiographic contrast medium containing naturally occurring radionuclides of thorium, and (2) low-LET radiations on animals. From these data, the cancer risk of low-LET radiation in humans was estimated. The liver cancer risk for alpha emitters was calculated to be 560 × 95 cases 10^{-4} Gy^{-1} (NCRP, 2001a). Extrapolation from animal data makes it possible to estimate the risk for human liver cancer from protracted exposures to beta/gamma emitters as 15 to 40 cases × 10^{-4} Gy^{-1}. These extrapolations were based on a linear, nonthreshold model. It was concluded that the liver is not a radioresistant organ.

Any attempt to extrapolate risk estimates across species must take into account not only the similarities and differences in the target cells and the initial events, but also the very important so-called host factors. These would include the factors influencing expression of the initial events such as hormones, cytokines, cell-cell communication, and the network of complex interactions. Not only the parenchymal cells are of importance, but also the stroma and the microenvironment. There has been a renewed interest in these aspects (Barcellos-Hoff and Brooks, 2001; Bissell and Barcellos-Hoff, 1987). However, the comparative differences in these elements, with the exception of hormones, such as the species-dependent differences related to breast cancer, have not been well studied. Thus, in this Section, the emphasis is on the cells of origin and the importance of comparing the responses not only on the basis of type of organ, but also on the same type of cell.

The types of cancer that are discussed in this Section have been selected because there is information, albeit incomplete, about the risk estimates of induction of cancer by radiation in humans and in at least one species of experimental animals. The degree of detail about the various tissues and the mechanisms of carcinogenesis vary markedly in the various parts of this Section. For example, the section on the hematopoietic system is much more detailed than the sections on solid tissues. The reason for this difference in detail includes the fact that there are data for the induction of various types of leukemia and lymphoma by radiation in humans and experimental animals, and there is increasing understanding of the mechanisms involved.

4.2 Hematopoietic System

4.2.1 Introduction: Leukemias and Lymphomas

In humans, leukemias and lymphomas of childhood are usually discussed separately from those in adulthood. Although there is not an absolute division in the types of leukemia that occur at different ages, there is a difference in the frequency with which certain types occur in childhood.

In the United States, acute leukemia is the most common malignancy among children. Acute lymphocytic leukemia (ALL) makes up ~75 % of the cases, with acute nonlymphocytic leukemias (ANLL), accounting for the rest of the cases (Poplack *et al.*, 1989). About 5 % of all cases of leukemia are myelogenous leukemias, which occur in two forms, one of which is acute (AML), the other is chronic myelogenous leukemia (CML).

The peak incidence of ALL occurs between 2 and 6 y of age. There is no clear-cut peak incidence of ANLL in relation to age. The median age of onset of the juvenile form of CML is at ~2 y of age, and the Philadelphia chromosome-positive type appears more frequently in older children. Among the atomic-bomb survivors, ALL was more common in children. An important feature of leukemia in the atomic-bomb survivors is that CML is found more frequently in both the control and exposed population in Hiroshima than in Nagasaki.

Acute leukemia occurs about 15 times more frequently in children with trisomy of chromosome 21 than children of the general population. The leukemias are usually ALL, but ANLL may occur.

The cells of origin in ALL have been determined through use of cell markers. Twenty to 30 % have T-lymphoblasts. In 1 to 2 %, the lymphoblasts have B-cell characteristics. The majority has no T- or B-cell markers and are called null cell ALL.

Of the cases of ANLL, between 50 to 70 % are classified as AML and 20 to 40 % as acute myelomonocytic leukemia. Childhood non-Hodgkin's lymphoma can be divided into two major immunological groups, lymphoblastic and nonlymphoblastic lymphomas. These types of leukemias and lymphomas may occur at older ages, but Hodgkin's disease and non-Hodgkin's lymphomas are the common types in adulthood.

The types of leukemia vary markedly in incidence among mouse strains. Myeloid leukemia occurs spontaneously at a low incidence rate in a few mouse strains, such as CBA/H, RFM, C3H/He, and SJL, and can be induced by radiation. Males are considerably more susceptible than females (Hayata, 1983; Mole et al., 1983; Ullrich and Preston, 1987).

The occurrence and susceptibility for lymphocytic leukemias in mice are likewise strain dependent. The dominant type occurring early in life is the disease characterized by an enlarged thymus involving immature lymphoblastic cells (Kaplan, 1974). The predominant cell type is the medium-to-large sized lymphocyte with the Thy-1 surface antigen and also a heterogeneous display of the differentiation markers common to T-cell subpopulations such as CD4 and CD8. Many of the studies of the induction of thymic lymphoma by radiation have been on the C57BL (Kaplan, 1974) and RF (Upton, 1977) strains. In contrast to myeloid leukemia, the preponderance of lymphomas occurs in female mice. There has been considerable refinement in the classification of lymphomas in the mouse through use of immunoglobulin and T-cell receptor probes (Vasmel et al., 1987). In studies of different F1 crosses, five suppressor regions on chromosome 4 have been putatively mapped (Herranz *et al.*, 1999; Melendez *et al.*, 1999). It has been shown that there is a loss of heterozygosity and that the regions showing such a change depend on the genetic background of the mouse (Meijne *et al.*, 2001). The results indicate the importance of a suppressor gene on proximal chromosome 4, and a possible role for *Tgfbr1* and *Pax5* genes (Meijne *et al.*, 2001). The fact that at least 13 loci on nine chromosomes have been associated with lymphomas illustrates the complexity of determining the targets and the alterations in the development of lymphoma (Boulton *et al.*, 2002). These authors found in their study of x irradiation of CBA/H × C57BL/Lia F1 hybrid mice that the thymic lymphomas were comprised of at least three distinct malignancies.

Late-life nonthymic lymphomas can occur in many strains of mice. In C57BL/6, the incidence rises in the second year of life. The precise phenotype of the lymphoma cells is not certain but is thought to be of the B-cell type. In many strains of mice, reticulum cell sarcoma occurs, but does not appear to be susceptible to induction by radiation. These reticular tissue tumors are divided into three types based on the histology and distribution of the neoplasia. Reticulum-cell neoplasm Type B resembles Hodgkin's disease in humans (Kaplan, 1974). The neoplasms of the reticular tissue in mice have been described in full by Dunn (1954).

In the chronically irradiated dog, myeloproliferative disease can reach a high incidence (Seed *et al.*, 1988), and the description of the development of acute myeloid leukemia in humans who have received radiation therapy matches that of the canine disease in many aspects (Fritz *et al.*, 1985; Tolle *et al.*, 1982; 1997).

4.2.2 Comparison of Radiation-Induced Leukemias Among Species

Three major human leukemic subtypes (*i.e.*, AML, CML, ALL) have been clearly defined as "radiation-inducible" hematopoietic neoplasias after single, acute, whole-body radiation exposures (0 to 4 Gy) (Preston *et al.*, 1994). Each of the three leukemic subtypes has reasonably close experimental counterparts based on common induction patterns, morphological and pathological characteristics; *e.g.*, for ALL, the C57BL/6 and AKR mouse strains (Kaplan, 1974; Upton, 1977; Upton *et al.*, 1958) and for myeloid leukemia, the CBA and RFM mouse strains or beagles (Mole *et al.*, 1983; Tolle *et al.*, 1982; Upton *et al.*, 1958). Absolute time and frequency parameters are, however, very different. Review of the various rodent and canine leukemic subtypes indicates that they can be readily fitted into the French-American-British classification scheme designed for human leukemias (Bennett *et al.*, 1976; Fritz *et al.*, 1985; Hayata, 1983; Tolle *et al.*, 1997). Specifically in terms of the myeloid leukemias, the M1, M2, M4, and M5 subtypes seem to predominate in humans and experimental animals (Fritz *et al.*, 1985; Hayata, 1983; LeBeau, 1992). By contrast, however, the M3 subtype, a promyelocytic leukemia, is rarely seen in humans or canines, but appears to occur more frequently in rodents.

4.2.3 *Pathology and Dose-Response Relationships*

The nature and temporal patterns of pathologic development for the various leukemic subtypes within irradiated mice, dogs and humans can often be strikingly similar. The description of the therapeutic x-ray induced acute myeloid leukemias in humans could well be that of the myeloid leukemias in chronically irradiated dogs, or acutely irradiated mice (LeBeau, 1992; Ludwig *et al.*, 1968; Rowley, 1985; Seed *et al.*, 1988; Upton, 1977). The evolving disease is first recognized by evidence of acute marrow suppression and pancytopenia, followed by a hypercellular myelodysplastic period in which all three major hematopoietic cell lines (erythroid, myeloid and megakaryocytic) are affected. With time and intensification of the myelodysplastic process, and in about one-half of the patients, a single fully transformed lineage finally emerges, culminating in AML (LeBeau, 1992), while in the remaining patients the myelodysplastic syndrome languishes without progression to AML, only for the patients to die of either infections or hemorrhage stemming from the pancytopenia. The course (prepatent/patent periods) of the therapy-related AML's is ~5 y (LeBeau, 1992). This compares with an induction time of ~3 y in chronically irradiated dogs (Seed *et al.*, 1988; Tolle *et al.*, 1997) and 1 y or less in acutely irradiated mice (Upton, 1977). In general, for relatively low doses (0 to 3 Gy) of either low- or high-LET radiations, the dose responses of the myeloid leukemias, AML and CML, have been adequately fitted with either simple nonthreshold linear, linear-quadratic, or quadratic-type models, commonly adjusted for a dose-dependent inactivation of targeted progenitors. For example, linear dose-response relationships have been used to describe developing myelogenous leukemia in adult male RFM mice acutely irradiated with either gamma rays (0 to 3 Gy; $Y = 0.67 + 0.065X$), or fission neutrons (0 to 0.8 Gy; $Y = 0.94 + 0.18X$), where Y is the percent tumor incident and X is the dose in rads (Ullrich and Preston, 1987). Other workers have analyzed related data sets for myelogenous leukemia induction by both high- and low-LET radiation in CBA and RFM mice (Barendsen, 1978; Mole, 1984; Ullrich and Preston, 1987) and have utilized linear-quadratic or quadratic equations, adjusted for target cell inactivation (Barendsen, 1978; Mole *et al.*, 1983). Barendsen (1978) used the linear-quadratic model to describe the dose-response curve for myelogenous leukemia in the RFM mouse, and Mole (1984) described the response in CBA/H mice as quadratic. Both authors corrected for cell inactivation but did not use data corrected for competing risks.

The radiation-induced leukemias in the atomic-bomb survivors and the associated risks have been adequately defined by several different nonlinear, nonthreshold-type models, including both linear-quadratic and linear-spline models (Preston *et al.*, 1994). However, when dose-response patterns for specific subtypes of leukemia were analyzed, distinct differences were noted: (1) for ALL, a linear-spline model provided the best fit of the data, although a threshold-type model could not be statistically eliminated; the dose-dependency for ALL incidence was highly significant, as was age-dependency (age at time of exposure); (2) for AML, the rise in incidence was highly dependent on both radiation dose and age-of-exposure, and best fitted by a nonthreshold, linear-spline response model; and (3) for CML, incidence was shown to be significantly dependent on the sex and exposure level of the individual, but not on age at exposure. The dose response for CML was best described by a simple, nonthreshold linear model (Preston *et al.*, 1994).

4.2.4 *Comparison of Hematopoietic Systems*

The basic structural design and the characteristics of cell proliferation of the hematopoietic tissue are similar in humans and the experimental animals that have been studied. The active bone marrow is in islands or cords supported by a complex stromal cell matrix that provides cytokines that regulate progenitor cell proliferation and other aspects of progenitor cell function. There is a common overall design of the proliferation in the bone marrow of the various mammalian species. There exits a stem-cell compartment from which daughter cells are fed into a compartment where amplification divisions occur, and differentiation increases as the capacity for proliferation is lost. The hematopoietic unit can be considered to consist of three elements: progenitor cells, stroma and the hematopoietins involved in regulation (Figure 4.1).

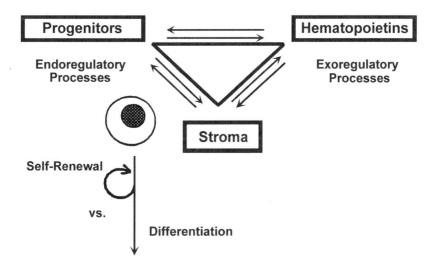

Fig. 4.1. Schematic of the relationship of the stroma, hematopoietins and the progenitors of hematopoietic cells.

4.2.5 *Target Cells*

The bone marrow consists of cells at all stages of development and of mature cells of the different lineages. The question is what cells are at risk for induction of the various types of cytokine network that has been altered by irradiation (Boniver *et al.*, 1990; Greenberger, 1991; Silver *et al.*, 1987; 1988).

These events are illustrated by: (1) the indirect leukemogenic action of stroma irradiation (both *in vivo* and *in vitro*) in eliciting growth factor-independent clones, and (2) the subsequent neoplastic potential within transplanted nonirradiated, growth factor-dependent progenitors (FDC-P1 cells) (Duhrsen and Metcalf, 1990; Greenberger *et al.*, 1990). The second observation points to the radiation-targeting of lineage-committed progenitors, thymocytes (Thy-1 + CD4 + CD8 + positive cells), suggesting a targeting specificity along the intrathymic T-cell differentiation pathway (Boniver *et al.*, 1990). Examples of retroviral-specificity and targeting of both specific cell lineages and, in turn, lineage specific leukemias, can be found in the literature (*e.g.*, Righi *et al.*, 1991; Watson *et al.*, 1987).

In summary, the concept of a time-based sequence of leukemogenesis common to all susceptible species is based on:

1. a direct radiation-induced initial event at the gene level of a primitive multipotential progenitor type (yet to be identified, but probably akin to a marrow repopulating, preCFUs-type stem cell); followed by
2. an indirect leukemogenic effect resulting from irradiation of the stroma that causes secondary events that influence the development of leukemia in the committed progenitors, such as GM-CFU, CFU-E, and BRU-E.

Therefore, by restricting the selective targeting of the post-initiation events of leukemia within lineage-committed progenitors, the expression of distinct leukemic subtypes in the diseased individual can be accounted for adequately (Figure 4.2).

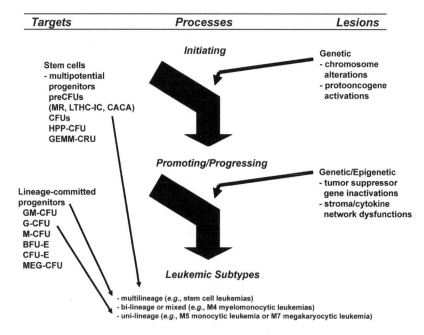

Fig. 4.2. A schematic of the targets, processes and lesions that occur in stem cells in leukemogenesis.

4.2.6 *Comparison of Cytogenetic Processes: Common or Species-Specific Patterns*

There are common and species-specific mechanisms for the major steps in the process of leukemogenesis. In terms of the "initiating" process by radiation exposure, several distinct types of induced cytogenetic lesions have been described and implicated. Due to unique karyotypes of the various species, the finding of unique, species-specific chromosomal lesions during the early and late prepatent periods of evolving leukemias is not surprising. Despite the numerical and spatial disparity of species-specific lesions, common mechanisms of induction of such lesions seem to exist. For example, for radiation-induced AMLs in various strains of mice, relative to those seen in humans and perhaps in dogs, the suspected initial lesion shares a number of common features, *i.e.*, high-frequency, nonrandomly occurring interstitial deletion/translocation-type lesion(s) localized to the "*q*" arm of chromosomes 2 and 5 of irradiated mice and humans, respectively (Rowley, 1985; Silver *et al.*, 1987), and probably chromosome 1 in dogs (Seed *et al.*, 1988).

The high-frequency nature of the above nonrandom cytogenetic lesions has been clearly documented in radiation-induced AMLs of various mouse strains; *e.g.*, in two studies, one by Hayata *et al.* (1983) and later by the Cox group (Breckon *et al.*, 1991; Silver *et al.*, 1987), the common chromosome 2 [del (deletion)] lesion was found in ~98 % (41/42) of the C3H/He AMLs; ~83 % (5/6) in RFM AMLs; and ~96 % (51/53) in CBA/H AMLs in acutely irradiated (x or gamma rays, C3H/He) mice. A third study by the Resnitzky group (Trakhtenbrot *et al.*, 1988) has also reported finding a common AML-associated chromosome 2 (del) type lesion in x-irradiated, corticosteroid-treated SJL/J mice. Additionally, when a few rare cases of spontaneous AML in nonirradiated C3H/He and RFM mice were examined, comparable chromosome 2 lesions were again noted (Hayata, 1983). The most striking feature of these chromosome 2 deletions and rearrangements was the commonality of the lesion pattern and its nonrandomness [*i.e.*, being largely confined to one chromosome 2 within regions C and D in the C3H/He and RFM AMLs (Hayata, 1983), and to regions C2 and E5-F in the AMLs of CBA/H mice (Breckon *et al.*, 1991)].

Comparable chromosome 2 type lesions within multipotential hematopoietic progenitors (of the preCFUs/marrow repopulating type) have been generated through *in vitro* irradiations and subsequent transplantation into lethally irradiated recipients (Silver *et al.*, 1987). Detailed cytogenetic analyses indicate that a number of these induced-lesions have a high degree of concordance not only with previously identified chromosomal fragile sites for clastogenic agents, but also for the above mentioned AML-associated lesions. The chromosome 2 sites in common are C2, F1, and F2 (Breckon *et al.*, 1991).

It is thought that two fragile recombinational domains are involved in the genesis of the deletion associated with AML in CBA/H mice (Bouffler *et al.*, 1993) and interstitial telomere-like repeat sequence arrays have been located. The molecular and cytogenetic findings (Bouffler *et al.*, 1997) indicate a significant interaction between the fragile domains after irradiation. The site of the putative tumor-suppressor gene has been mapped to a 1 cM segment on mouse chromosome 2 using microsatellite-based allelotyping of F1 hybrid mice (female CBA/H × male C57BL/Lia) (Silver *et al.*, 1999). More recently, Finnon *et al.* (2002) suggested that the domain-specific fragility of chromosome 2 associated with AML is in the domains rich in gene-related products essential to hematopoeisis. It is clear that the strand-dependent, and easily identified, aberration of chromosome 2 is associated with complex, and, as yet, not fully understood molecular changes.

Related high-frequency, nonrandom deletion-type lesions, in specific chromosomal regions 5q- and 7q-, have been reported as being associated with the radiation or chemotherapy-induced AMLs and myelodysplastic syndromes of humans (Rowley, 1985), as well as with acute irradiation-induced AMLs of the atomic-bomb survivors (Kamada, 1991). Critical chromosome regions, *i.e.*, 5q23-q32 and 7q32-q35, have been identified, with selected regions overlapping with known fragile sites for selected oncogenic/ clastogenic agents (Djalali *et al.*, 1987; LeBeau, 1992). Hematopoietic elements bearing these lesions first appear shortly after therapy, clones subsequently evolve displaying the severity of the myelodsyplastic syndromes, and then AML develops (Rowley, 1985). By contrast, in the radiation-associated CMLs of humans, the frequencies of the pathognomonic cytogenetic lesion [*i.e.*, the Ph1 t (9, 21) chromosome] remains proportional to the incidence of the disease, not disproportional with elevated radiation exposures (Kamada, 1991). Paradoxically, this observation seems to argue against the prevailing concept that radiation initiates CML in humans *via* the induction of the Ph1-type lesion.

Experimental homologs of the Ph1-like cytogenetic lesion have been reported in several different experimental animal models, although these lesions are not nearly as well defined as the chromosome 2 (del) AML-associated lesions in irradiated mice (Breckon et al., 1991; Hayata, 1983). For example, in canines, a small, acrocentric, Ph1-like marker chromosome has been observed in ^{90}Sr-induced myelomonocytic leukemias (Shifrine et al., 1971). In alpha-particle irradiated mice (C57BL/6), a marker chromosome derived from a chromosomes 1:5 translocation, has been detected during early preclinical phases of evolving thymic leukemia/ lymphoma and occurs in proportion to the frequency of preneoplastic thymic cells (McMorrow et al., 1988).

In terms of the relationships of the genes involved, it is currently recognized that the actively transforming, radiation-activated oncogene, v-*abl*, is derived from a comparable Ph1-like gene fusion that involves the fusion of retroviral *gag* sequences with the third exon of c-*Abl* (as compared to the *BCR/ABL* gene fusions found in Ph1-positive CMLs and ALLs of humans) (Rosenberg and Witte, 1988). Despite these differences in molecular construction of the hybrid genes, the final gene products, such as *p210* and *p185*, in all cases appear to be similarly endowed with elevated tyrosine kinase activity (Sawyers, 1992).

4.2.7 Leukemogenesis Resulting from Gene Rearrangements

When human CML *BCR/ABL* hybrid genes are expressed in either transgenic or in transplanted mice (with hematopoietic cells bearing transfective human *BCR/ABL* genes), a wide array of hematopoietic disorders is induced, including not only a prominent CML-like expansion of both precursor and mature neutrophilic compartments of the marrow and blood (Daley et al., 1990), but also of other B-cell leukemias/lymphomas, erythroleukemias, and reticulum cell sarcomas of macrophage or histiocytic origins (Elefanty et al., 1990).

Further, more direct support for the prevailing concept that the hybrid *BCR/ABL* gene and related homologs can initiate leukemia and cause a dysfunctional hematopoietic expansion, has been obtained by several novel *in vitro* experiments [*e.g.*, retroviral-mediated transfection of the *BCR/ABL* gene into targeted multipotential stem cells and subsequent observation of deregulated clonal functions, including markedly enhanced clonality, along with reduced growth-factor dependence, but with continued maintenance of apparently normal capacity for multilineage differentiation, and the absence of overt signs of neoplastic conversion (Gishizky and Witte, 1992)].

4.2.8 *Secondary Cytogenetic Lesions Associated with Leukemia Promotion and Progression*

What is known about the cytogenetics of leukemia of the so-called promotion and progression stages of evolving radiation-induced leukemia is meager by comparison to the suspected initiating lesions described above. Nevertheless, several distinct types of secondary cytogenetic alterations have been observed and implicated in the promotion and progression of late stages of various leukemic subtypes within various species. For example, the radiation-induced AMLs of CBA/H mice commonly exhibit, but at moderately low frequency (~15 %), trisomies of chromosomes 1, 5, 6 and 15 and a monosomy of chromosome Y (Breckon *et al.*, 1991). Comparable trisomies (chromosome 15) are commonly detected during late developmental stages of radiation-induced thymic lymphomas in chronically irradiated C57BL mice (Janowski *et al.*, 1990). Although consistently noted, these late arising chromosomal duplications occur at moderately low frequencies (~15 %) (Breckon *et al.*, 1991). Similarly, in humans, chromosome 8 duplications are common, but occur at moderately low frequency (~16 %), not only in the therapy-induced secondary AMLs (Rowley, 1985), but also in the blast transition phase of evolving CML (Rowley and Testa, 1982). It should be noted, however, that at least two other common cytogenetic events (isochromosome 17q, a second Ph1 chromosome) also seemingly accompany blast transition in CML.

Regardless of the structural nature or location of these putative "critical" late-stage cytogenetic lesions, they must by definition effectively enhance survival and selection of the lesion-bearing progenitors, and, in turn, promote an unbridled clonal expansion. With the current recognition that selected types of protooncogenes and tumor-suppressor genes cooperatively regulating cell growth, it is likely that the above cytogenetic lesions are indeed causally linked to late-leukemogenic stages by way of selectively- and cooperatively-altered oncogenes and tumor-suppressor genes. For example, the chromosome 8 trisomy in the radiation therapy-induced AMLs and CMLs in humans might provide the requisite clonal advantage through an additive, *MYC*-mediated gene dosage effect due to chromosome 8/*MYC* duplication on the initiated cell's proliferative capacity, whereas the i(17q) lesion might cooperate in a subtractive mode *via* an induced loss of growth regulating functions imparted by the *TP53* tumor-suppressor gene, known to be localized in the q arm of chromosome 17.

In summary, the high degrees of concordance between the site-specific and structural features of high-frequency, nonrandom, interstitial deletion/translocation-type chromosomal lesions during the early preleukemic period compared with those in the latent period of evolving AML lend support to the concept that these are "initiating" lesions of leukemia by nature.

By contrast, however, in terms of radiation-induced CMLs, there is little direct support of the prevailing notion that radiation initiates the leukemic process by a direct induction of a Ph1-like lesion. However, there is some evidence indicating that critical leukemia initiating lesions actually arise from radiation-targeted fragile sites that are parentally-based and genomically imprinted (Breckon *et al.*, 1991), and that these initial lesions are the precursors of the more complex gene rearrangements noted above. Nevertheless, in the absence of essential secondary and tertiary processing steps by the hematopoietic microenvironment, these "initiated" progenitors consistently fail to evolve clonally into the fully transformed leukemic cell phenotype (Silver *et al.*, 1987).

4.2.9 *Hematopoietic Cell Origins of the Putative "Critical" Genic Lesions and the Nature of Induced Genic Dysfunctions*

In the case of AML in humans associated with radiotherapy, the deletion-type 5q-lesions have been directly linked to loss of a battery of cytokine and cell regulatory genes, and, in turn, a general breakdown of growth controls. Some of the gene clusters that are lost or rearranged include: *IL3, CSF2, IL4, IL5, TCF7, GMF1, CD14, FGF1, NR3C1, SPARC, ADRA1* and *ADRB2* (LeBeau, 1992).

By contrast to the *IL1b* alterations associated with the chromosome 2 (del) lesion, a second known murine chromosome 2-located gene, the protooncogene *ABL* (and a key gene element in human CMLs) appears not to be rearranged, amplified or overexpressed as a result of radiation-induction of chromosome 2 (del) lesion (Silver *et al.*, 1987). This apparent lack of the putative leukemia-initiating *ABL* gene's early involvement in the development of these acute radiation-induced myeloid leukemias in mice, tends to argue against a kinship to CML in blast crisis, while favoring by default a relatedness to human AML.

The above situation is probably not the case with the chronic radiation-induced canine myeloid leukemias, as molecular analyses of the *ABL* protooncogene have revealed genic rearrangements and overexpression (Frazier *et al.*, 1988). However, in mouse and dog, definitive proof of a radiation-induced, translocation that results in a hybrid gene product, comparable to *BCR/ABL p210* in humans, has yet to be detected.

Nevertheless, it should be again noted that the hybrid human CML *BCR/ABL* gene and its gene product, *p210*, can elicit CML-like disease in mice following either gene transfection or direct injection of *p210* (Daley *et al.*, 1990; Heisterkamp *et al.*, 1990). Further, the experimental induction of a Ph1-positive, ALL-like human disease in experimental mice has been obtained by transfecting a related hybrid *BCR/ABL* gene (containing only *BCR* exon number 1, rather than exon number 2) and eliciting expression of its transforming gene product, *p190* (Hariharan *et al.*, 1989; Heisterkamp *et al.*, 1990). The genetically engineered human ALL in mice appears closely related to the naturally occurring or mutagen-induced murine lymphoid leukemias and lymphomas, in that the viral oncogene, v-*abl*, is also a hybrid gene, formed by the fusion of the retroviral *gag* sequences with exon number 3 of c-*Abl* (Sawyers, 1992).

In all of the above human and murine CML and ALL diseases, the transforming activity of the hybrid gene fusion relates to the activation of the *Abl* tyrosine kinases (Sawyers, 1992). Further, all of the gene products can seemingly induce experimentally marked myeloid expansions, comparable to that seen in human CML during its chronic phase (Sawyers, 1992).

4.2.10 *Cooperating Oncogenes in Lymphoid Neoplasias*

One of the more significant biological discoveries over the past several decades concerned the nature and cellular origins of the cancer related genes, the so-called oncogenes. For many years it has been well recognized that ionizing radiation, along with a host of other toxic carcinogenic agents, can activate latent retroviruses that contain transforming gene elements that, as it turns out, were originally derived from normal but mutated and rearranged cellular genes of the host cell (Aaronson and Tronick, 1985; Temin, 1985). More recently, it has become clear that a number of these transposed activated oncogenes cooperate in the transformation process within specific target cells and in the evolution of type-specific neoplasias (Adams and Cory, 1992; Hunter, 1991).

In both T- and B-lymphoid elements during preneoplastic periods, the activation or over expression of the oncogene c-*Myc* is closely tied to enhanced cell proliferation, with curtailed maturation, but without the induction of growth autonomy (growth factor independence) (Langdon *et al.*, 1986). However, when c-*Myc* gene functions are combined with the activities of a number of other oncogenes (*Ras, Raf*), tumor-suppressor genes (*i.e., Bcl2, Trp53*), or normal cytokine genes (*Il6*), full transformation is commonly seen

(Adams and Cory, 1992). For example, transgenic mice bearing activated *Myc* and N*Ras*, rapidly develop pre-B- and pro-B-type leukemias/lymphomas (Harris *et al.*, 1991). The activated oncogene, *Abl*, appears to cooperate similarly with *Myc* in the transformation process, but targets a more differentiated stage of B-cell development (plasma cell). The cooperativity between the tumor-suppressor-like gene, *Pbbcp2*, and *Myc* seemingly enhances clonal longevity of targeted B-cell progenitors and eventually elicits their immortalization.

Comparable cooperative gene functions of *Myc* and several other oncogenic loci (*Ras, Pim1, Bmi1*) within T-cell/pre-T-cell targets have been associated with rapidly developing T-cell type lymphomas/leukemias in neonatal mice (Haupt *et al.*, 1992; van Lohuizen *et al.*, 1989).

4.2.11 *Cooperating Oncogenes in Myeloid Neoplasias*

Like the lymphoid lineages, several myeloid lineages are also susceptible to leukemic transformation following cooperative interaction of a number of different gene classes; *e.g.*, the proliferation-associated oncogene *Myc* functions in combination with a number of other oncogenes (*Ras/Raf*), as well as with several different genes that produce cytokines such as *Il3*, CSF1, transcription factors (*spi-1, fli-1*) or homeobox genes (*Hoxb8*) (Perkins *et al.*, 1990). A paradigm of the basic leukemogenic process is provided by gene analysis of WEHI-3B cells, a myelomonocytic-type leukemia, and its altered patterns of gene expression of two vital genes, *Il3* and *Hoxb8*. The induction of a constitutive genic expression of *Il3* provides for a continuous source of essential growth-factor, while the modified *Hoxb8* expression provides for the retardation of any differentiative functions normally imparted by *Il3*. The cooperative role played by each of the latter genes (*Il3* and *Hoxb8*) has been experimentally demonstrated by retrovirus-mediated transfection of individual genes into suspectible myeloid progenitors. Transfection of *Il3* alone results in an excessive, nonneoplastic proliferation, whereas the *Hoxb8* gene alone results simply in a reduced differentiative flow. However, together the two transfected genes actively transform targeted myeloid progenitors (Perkins *et al.*, 1990; Perkins and Cory, 1993).

In terms of the putative "initiating" chromosome 2 (del) lesion found in radiation-induced murine AMLs, comparable cooperative interactions have been found between the overexpressed *Il1b* gene and the loss of function (single copy) of the *Hoxd3* gene (Cleary, 1992).

Although the above examples relate largely to spontaneous murine leukemias, similar types of gene cooperation seemingly play a role in the genesis of human leukemias (*e.g.*, *PBX* and *TLX1* homeobox genes) (Cleary, 1992). Also, under select exposure conditions, ionizing radiation apparently can precipitate these cooperative gene responses as well, although the initial triggering events remain elusive.

4.2.12 *Hematopoeitic Microenvironment*

It is not only the cell involved in the initial events leading to leukemias that may be species-specific because there may also be differences in the microenvironments. For example, cells exist in different environments consisting of extracellular matrix proteins, cytokines, soluble hormones, and cells of other types. In the case of murine thymic lymphoma, the thymic microenvironment plays a role in the development of normal thymocytes and in the development of thymic lymphomas. For example, stromal "nurse" cells provide signals for differentiation of thymic progenitor cell subsets and disruption of this process can facilitate the development of thymic lymphoma. [For discussion of the role of the microenvironment on thymic lymphoma, see Boniver *et al.* (1990), Muto *et al.* (1991), Sado *et al.* (1991)].

Comparable stromal factors involved in myelopoiesis and disordered function of the stroma and associated cytokine network are thought to influence the development of myeloid neoplasias. There is not sufficient information about similarities and differences among species in either the nature or the functions of the stroma to know whether it is an important consideration in the extrapolation of risk.

The acquisition of autocrine functions can give initiated hematopoeitic clones a growth advantage. Regardless of the species involved, the preleukemic period appears to be associated with a time-dependent degradation of the stroma-cytokine network that regulates the growth of clones with altered growth and phenotypes (Greenberger *et al.*, 1990). Normally, two processes, differentiation and proliferation that are influenced by cytokines, are held in balance. A loss of this balance can result in the growth of initiated cells (Seed, 1991). A number of relevant findings have been reported. For example, autocrine production of selected cytokines appears to be a critical predisposing leukemogenic event in that autocrine production of an essential cytokine (GM-CSF) has been observed in ALL, AML and CML. Further, autocrine production has been induced by *in vitro* transfection of *BCR/ABL* into susceptible cell lines (Elefanty *et al.*, 1990). Murine HSCs transfected with *BCR/ABL* constructs and subsequently transplanted into lethally irradiated mice develop several different hematopoietic diseases. Finally, the *BCR/ABL* transfected construct confers proliferative advantage on diverse hematopoietic cells, but complete transformation requires additional genetic change.

4.2.13 *Summary*

Both the myeloid and lymphoid systems are susceptible to neoplastic change. There are two types of the neoplastic diseases of the myeloid component of the hematological system for which there are animal models and that have been studied extensively. First, myeloproliferative disease that occurs in the chronically irradiated beagle dog has an extensive database. This disease resembles what is seen in the development of acute myeloid leukemia in humans associated with prior exposure to radiation. The initial events involve chromosome aberrations, and secondary events are mediated by a microenvironment involving the stroma-cytokine network that is altered as a result of the irradiation. Second, in certain strains of mice a low background rate of acute myeloid leukemia is found and the males of these strains are quite susceptible to the induction of leukemia by radiation. A consistent characteristic finding in mice with leukemia is an aberration on chromosome 2. This is not the only aberration found and the complex molecular mechanism and how it compares with that in humans is being investigated.

The natural occurrence, the susceptibility for induction by radiation and the type of neoplasia of the lymphoid system are strain dependent in mice. Similar types of leukemia are also found in rats. The dominant type that occurs in young, and most frequently in female mice, involves immature lymphoblastic cells and is called thymic lymphoma. The clinical characteristics of leukemia involving similar types of cells are different in humans. In mice, about 13 loci on nine different loci have been implicated in the development of radiation-induced thymic lymphoma. Induction is most efficient when young female mice, particularly of the C57BL strain, are given about four weekly treatments of ~1 Gy. Later in life nonthymic lymphomas occur, and in some strains the incidence is high. One type of the nonthymic lymphoma, reticulum cell Type B, is similar to Hodgkin's disease in humans. There has been no success, so far, in extrapolating risk estimates of leukemias in experimental animals to humans. Even attempts to extrapolate from one strain of mouse to another failed (Storer *et al.*, 1988). However, there is hope that the mechanism of induction of acute myeloid leukemia in both humans and mice will be elucidated and that may make it possible to develop an approach to extrapolation.

4.3 Lung

4.3.1 *Introduction*

The major histological types of lung cancer in humans are adenocarcinoma, small-cell carcinoma, SCC, and large-cell carcinoma. With respect to their distinctly different responses to cytotoxic therapy, human lung cancers are frequently classified into two broad categories of: (1) SCLC and (2) nonsmall-cell carcinoma (Seifter and Ihde, 1988). However, in experimental lung cancer research aimed at elucidating mechanisms of initiation and progression of the disease, it is well advised to make every effort to subclassify in as much detail as possible the experimental tumor system under study. Likewise, novel strategies for lung cancer prevention and therapy, many of which attempt to target mitogenic signal transduction pathways, require a more detailed characterization of the tumors to be treated as many pathways are cell type specific (Schuller, 1991).

Although the majority of the histopathological lung cancer types seen in human patients do have counterparts in either spontaneous or experimentally-induced lung tumors in animals, the origin and development of these tumors in the animal models do not in all cases mimic human disease. Admittedly, it is not possible to conduct true pathogenesis studies in humans, but a large body of evidence has accumulated over decades describing various stages of preneoplastic changes either simultaneously with the occurrence of fully developed tumors or in identified high-risk populations prior to the development of tumors. Moreover, the anatomical localization of the tumors, for example, whether they are central or peripheral and whether they involve bronchi, bronchioles, alveoli or subbronchial glands is often different in rodents as compared with humans.

Of the many different cell types found in the lung, only a few have been implicated in the development of lung cancer, but the anatomical location of the possible cells of origin of the cancers differs among species (Keehn *et al.*, 1994; Land *et al.*, 1993a; Plopper *et al.*, 1980; Reznik-Schuller and Reznik, 1979; Saccomanno *et al.*,

1996). Since there are significant differences between the types of lung cancer that are important when extrapolation of data is being made across species, the major types of cancers are considered separately. Lung cancers induced by internally-deposited radionuclides, such as plutonium, are discussed in Section 6.6.5.3.

4.3.2 *Adenocarcinoma*

The incidence of pulmonary adenocarcinoma has increased in males and females in the United States in the last two decades and is now the most common type of lung cancer in women (Devesa *et al.*, 1991; Linnoila, 1990). Smokers are at risk to develop this type of cancer; the incidence of adenocarcinomas is higher in nonsmokers than that of SCC and small-cell carcinoma (Linnoila, 1990; Wynder and Covey, 1987). Studies of families (Tokuhata and Lilienfeld, 1963) and relevant data from monozygotic twins (Joishy *et al.*, 1977) suggest that a genetic factor is involved in susceptibility. Adenocarcinomas arise in the lung periphery (Linnoila, 1990) and occasionally from epithelial cells of larger bronchi and their submucous gland (Shimosato, 1989). The tumors are gland-like with many Clara cells, alveolar type-II cells, or mucous cells (WHO, 1981). On electron microscopic examination, microvilli tight junctions and immature secretory products are seen (Schuller, 1989). For many years the cell of origin was thought to be alveolar type-II cells but based on electron microscopy, the majority of cases appear to arise from bronchiolar Clara cells (Albertine *et al.*, 1998; Linnoila, 1990).

Adenocarcinomas are the common type of lung cancer in mice and are characteristically peripheral in location. A distinction is often made between adenomas and adenocarcinomas. It is not clear whether adenomas are benign because most are transplantable and therefore meet one of the criteria of malignancy. Lung tumors in mice are often classified as alveologenic or bronchiolar. It is thought that the former arise from alveolar type-II cells and the latter from Clara cells. Therefore the cell type of origin of adenocarcinomas can be the same as in humans. Peripheral adenocarcinomas are produced in mice by a number of chemical agents such as nitrosamines (Belinsky and Anderson, 1990; Belinsky *et al.*, 1990; Malkinson, 1992; Reznik-Schuller and Reznik, 1979) and by radiation (Fry and Witschi, 1984).

In A/J mice, perhaps the most studied strain in relation to chemical agents, the tumors arise from alveolar type-II cells and express point mutations, in codons 12 and 61 of the K*Ras* gene (Belinsky *et al.*, 1996). The most common of these mutations in codon 12 is also the most common point mutation seen in human peripheral adenocarcinomas. However, most of human adenocarcinomas (Rodenhuis *et al.*, 1988) arise from Clara cells and not alveolar type-II cells.

The difference between alveologenic cell and Clara cell tumors may be important when considering extrapolation because studies suggest deficiencies in the cyclic-AMP dependent signal transduction pathway in murine pulmonary adenocarcinomas (Malkinson, 1992). This pathway has been found in human Clara cell tumor cell lines.

The peripheral adenocarcinoma in the rat has been used in the study of chemical carcinogenesis (Belinsky and Anderson, 1990; Belinsky *et al.*, 1990). In the rat no alterations in the *Trp53* gene were found in adenocarcinomas induced by a number of chemical carcinogens (Belinsky *et al.*, 1997). Although the tumors are not of the Clara cell type, it is in these cells that the DNA-nitrosamine adducts are found.

The alterations in the KRas gene and *p53* proteins in rat lungs were found to be influenced by the type of causative agent. For example, mutations in the KRas gene were found much more frequently after exposure to plutonium than gamma rays and were not found in naturally occurring tumors (Belinsky *et al.*, 1997). It is thought that mutations in the KRas gene are important in progression and not in the initial events (Dragani *et al.*, 1995).

In the Syrian golden hamster, peripheral adenocarcinomas arise from Clara cells after exposure to nitrosamines. A high incidence of point mutations in codons 12 and 61 has been found in these Clara cell tumors (Oreffo *et al.*, 1993), which is also the case in humans. Since this mutation is commonly found in murine alveologenic tumors, perhaps the molecular changes in the initial events are comparable for lung tumors arising from distinct cell types. At least this appears to be a change that is common in hamsters, mice and humans, but not in rats. Adenocarcinomas were the most common type of lung tumor in female atomic-bomb survivors.

4.3.3 *Squamous-Cell Carcinoma*

Squamous-cell carcinoma (SCC) used to be the most common type of human lung cancer. It has a strong association with cigarette smoking (Weiss, 1983), but has declined in incidence in the last two decades (Linnoila, 1990). It still accounts for ~30 % of all human lung tumors and the incidence is higher in males than females. The most common type of lung cancer in male atomic-bomb survivors was squamous cell (Land et al., 1993a), but the relative proportion of cases was not related to dose. This type of cancer is predominantly central in location and arises from the respiratory epithelium of the stem lobar and first order segmental bronchi (Linnoila, 1990; Melamed and Zaman, 1982). The distinction of SCCs from other types is based on evidence of squamous differentiation, such as keratin or its precursors (tonafilament bundles detected by electron microscopy) and/or the presence of prominent intercellular bridges, which are confirmed as desmosomes by electron microscopy (Melamed and Zaman, 1982; Schuller, 1989). The cell of origin is in dispute. For many years it was suggested that basal cells underwent hyperplasia, progressed to dysplasia and carcinoma *in situ*, thus implicating basal cells as the cell of origin. This is still held by some experts to be the pathogenic pathway. Others suggest that the tumors could arise from secretory cells (McDowell, 1987). It is quite possible that the pluripotent stem cell, which has not been identified but presumably belongs in the basal epithelium, will turn out to be the cell of origin (Linnoila, 1990). The question of cell of origin is important in understanding the mechanism of the induction of lung cancer by radon. The risk of lung cancer from exposure to radon is based entirely on epidemiological data (NAS/NRC, 1999). The risk might be greater if the more superficial secretory cells are at risk because the alpha particles from the radon progeny are of low energy and more likely to traverse the secretory cells than the basal cells. In human tumors, point mutations in members of the *RAS* family have been noted (Rosell et al., 1993). Mutations in the *TP53* tumor-suppressor gene, a common finding in many types of solid cancers, are also found. Another factor is the expression of abundant epidermal growth factor receptors (Kelly et al., 1991; Piyathiluke et al., 2002).

SCCs have been induced in mice and rats with chemical carcinogens by special techniques such as instillation. This type of tumor is not a common outcome of irradiation in experimental animals although peripheral squamous-cell tumors were reported after instillation of ^{210}Po. No animal model is very satisfactory for this type of tumor.

4.3.4 *Small-Cell Lung Carcinoma*

Small-cell carcinoma of the lung accounts for ~30 % of all human lung cancers (Linnoila, 1990; Weiss, 1983) and demonstrates a strong etiological link with smoking (Weiss, 1983). Epidemiological studies in uranium miners have provided evidence that simultaneous exposure to alpha particles and cigarette smoke greatly enhances the risk of developing this cancer type (Weiss, 1983; 1991). From studies of uranium miners and atomic-bomb survivors, radiation-induced lung cancers appear to be more likely small-cell carcinomas and less likely to be adenocarcinomas (Land et al., 1993a). More recent evidence also suggests that smokers who develop chronic obstructive lung disease may be at a particularly high risk to develop this cancer type (Weiss, 1991). Small-cell cancer may arise both centrally in association with large airways (~60 %) or in the lung periphery (~30 %). Attempts to subclassify this cancer type with respect to cell size and shape [e.g., oat cell versus intermediate cell as suggested by the World Health Organization (WHO) classification] are of little consequence as there is great heterogeneity within a given tumor. This cancer type typically expresses a variety of neuroendocrine markers such as the production of biogenic amines, neuropeptides, and/or calcitonin, chromogranin, neuron specific enolase, and L-dopa decarboxylase activity (Gazdar, 1984; Kelly et al., 1991; Marangos et al., 1982). Many of the neuropeptides (Gazdar, 1984) as well as the biogenic amine serotonin (Cattaneo et al., 1993; Schuller, 1989) are autocrine growth factors, thus enabling the tumor cells to stimulate continuously their proliferation.

In contrast to the highly malignant small-cell cancer, carcinoids are relatively benign lung tumors that express the same spectrum of neuroendocrine markers (Eggleston, 1984; Gazdar, 1984). Carcinoids are subclassified into typical and atypical carcinoids, the latter having a more varied cellular morphology and distinct invasive and metastatic potential (Eggleston, 1984). These tumors may remain silent for long periods of time and are usually only diagnosed if their neurosecretory products cause clinical symptoms such as Cushing syndrome. They have not been linked to any particular etiological agent thus far, possibly because of their rare occurrence.

Because small-cell cancers and carcinoids express morphological and functional features of neuroendocrine cells, it is widely believed that they are derived from pulmonary neuroendocrine cells (Becker and Gazdar, 1983; Bensch et al., 1968). Pulmonary neuroendocrine cells occur at all levels of the airways including central bronchi, peripheral bronchioles, and alveoli. However, they are sparse in healthy adult mammals, which makes it difficult to imagine that such a frequent cancer type as SCLC is derived from them.

Although SCLC and carcinoids express a similar spectrum of neuroendocrine markers, they differ quite significantly at the molecular level. SCLC typically demonstrates a high incidence of changes in *MYC* family oncogenes [overexpression, amplification (Nau *et al.*, 1986; Wong *et al.*, 1986) while lacking changes in *RAS* family oncogenes (Mitsudomi *et al.*, 1991)]. Carcinoids lack changes in *MYC*-family genes and demonstrate a significant incidence of point mutations in *RAS* family genes (Mitsudomi *et al.*, 1991). SCLS express a host of autocrine growth factor pathways, the majority of which activate protein kinase C downstream (Bunn *et al.*, 1992; Kelly *et al.*, 1991) whereas carcinoids, in addition to such pathways, express an EGF receptor-mediated mitogenic pathway that is not expressed in SCLC (Moody *et al.*, 1990). Both, SCLC (Cattaneo *et al.*, 1993; Maneckjee and Minna, 1990; Schuller, 1991) and carcinoids (Schuller, 1991) have been shown to express nicotinic acetylcholine receptors.

Studies of the spectrum of mutation in the *TP53* gene in lung tumors from uranium miners and smokers (Vahakangas *et al.*, 1992) suggested that the spectrum of mutations was different. Clusters of mutations in codons 141 to 161 and 195 to 208 were found in the uranium miners and interpreted as "hot spots" for mutations caused by the alpha particles from the radon progeny. However, Taylor *et al.* (1994) reported mutations in codon 249 as the possible marker of exposure to radon. The validity of those observations is discussed in Section 5.3.5.2.

An animal model for human SCLC has yet to be discovered as this histological tumor type has neither been experimentally induced nor reported as a spontaneous disease in any animal species to date. Experiments using polycyclic aromatic hydrocarbon and ferric oxide as cancer-inducing agents resulted in poorly differentiated malignant lung tumors of small, oat-cell like cells, suggesting that they may be small-cell cancers (Blair, 1979). Unfortunately, the diagnosis was not verified by special staining techniques (*e.g.*, silverstains), and immunocytochemical stains for neuroendocrine markers were not available at that time. It would seem worthwhile to repeat this study applying today's more advanced diagnostic tools.

Experiments with male Syrian golden hamsters have produced neuroendocrine lung tumors by simultaneously exposing the animals to hyperoxia and N-nitrosodiethylamine (Nylen *et al.*, 1990; Schuller *et al.*, 1988) or the tobacco-specific 4-methylnitrosamino-1-(3-pyridyl) butanone (Schuller *et al.*, 1990). These tumors arose in the lung periphery from foci of alveolar type-II cells that differentiated into a neuroendocrine phenotype as an early event of preneoplastic development (Nylen *et al.*, 1990; Schuller *et al.*, 1990). The tumors were classified as atypical carcinoids.

There are no experimental studies on radiation-induced neuroendocrine tumors.

4.3.5 Large-Cell Carcinoma

The term "large-cell carcinoma" is generally assigned to a lung tumor that does not demonstrate evidence for neuroendocrine, squamous or adenomatous differentiation with light microscopy examination (Linnoila, 1990). As many of these tumors demonstrate focal areas of differentiation suggestive of one of the three above cancer types or immature differentiation features only discernible by special techniques such as electron microscopy (Schuller, 1989), they may either be derived from better differentiated cancer types at a late stage, or grow too rapidly for cellular differentiation to become fully developed. Large-cell carcinomas are primarily peripherally located in the lung, and their biological behavior and response to therapy are somewhat between that for the other histological lung cancer types (Linnoila, 1990).

There is no animal model for large-cell carcinoma. Experiments in laboratory animals unequivocally result in the induction of fairly well differentiated lung tumors that are readily classified by the investigators. This may to a great extent be a reflection of the short life span of laboratory animals, which precludes the development of highly advanced, dedifferentiated tumors. Moreover, for experimental tumor systems, animals are generally exposed to one or very few agents aimed at inducing the cancer and/or modulating its development. Humans are exposed, however, to a multitude of endogenous and exogenous factors that may affect cellular responses such as proliferative ability, differentiation and metastasis. In support of this interpretation, Witschi and Schuller (1991) reported the induction of a significant incidence (42 %) of poorly differentiated, highly malignant tumors that, by lack of specific diagnostic features, justify classification as large-cell carcinomas, in male Syrian golden hamsters exposed simultaneously to hyperoxia and ^{210}Po. Unfortunately, this interesting model has not yet been further pursued.

4.3.6 Summary

Lung is a suitable organ for testing methods of extrapolation and identifying the problems and possibilities because there is a considerable amount of data for the naturally occurring and radiation-induced tumors in humans and rodents. Tumors of the lung vary in type and location among different species. For example, SCLC accounts for ~30 % of all lung tumors in humans but does not occur in rodents. It is important that attempts to extrapolate risk estimates of radiation-induced lung tumors are based on data for the same type of tumor and not on the total lung cancer incidence.

Adenocarcinoma is the most common type of lung cancer in mice and is characteristically located peripherally in humans and mice. Both the incidence of naturally occurring tumors and susceptibility to induction are strain dependent. There is also some evidence that there is a genetic basis of susceptibility for this type of lung cancer in humans. The tumor in humans and mice may arise from Clara cells or the relevant stem cell. SCCs do not appear to be induced by radiation in mice.

There is a need for a greater understanding of the cell of origin of the different types of tumors, the mechanism of the induction and the importance of differences in host factors before progress can be made in the task of extrapolating risk estimates for all lung cancers.

Unless a method is derived that eliminates the problem of extrapolating risk estimates based on the tissue site but not the cell or origin, approaches in extrapolation should concentrate on tumors of the same cell of origin.

4.4 Breast

4.4.1 *Histogenesis of Mammary Glands and Mammary Cancer*

Mice and rats have been used extensively in the study of mammary carcinogenesis and the induction of cancer by radiation. In the case of mice, the studies have involved the role of the murine mammary tumor virus, genetics, and molecular mechanisms, and because of the marked strain-dependent differences in susceptibility to naturally occurring and radiation-induced cancer. The ability to modify the genetic make-up has increased the potential for elucidating mechanisms of action. The rat is considered an appropriate overall model because growth and differentiation of the gland, and cancer development in female rats share features in common with those in women. Russo and Russo (1987) have analyzed the development of the mammary glands of line-bred Sprague-Dawley rats. The progressive arborization and growth of the developing gland into the surrounding fat in the young occurs *via* proliferation in the terminal end buds (TEB) at the tip of the duct. With the advent of estrous cycles at 35 to 42 d, TEB begin to differentiate progressively into alveolar buds, which, with pregnancy, differentiate into grape-like lobules of clustered fully differentiated alveoli. Some ducts do not do so; the cells at their tips are reduced in number and they become quiescent terminal ducts.

The proliferative activity and the stage of differentiation largely dictates sensitivity to chemical carcinogens. A window of high sensitivity to mammary cancer induction by the chemical carcinogens dimethylbenzanthracene (DMBA) and methyl-N-nitrosourea, for example, occurs during the period of active proliferation and differentiation of TEB into alveolar buds from the onset of estrous cycles until ~55 d of age in Sprague-Dawley rats (Huggins et al., 1959; Russo and Russo, 1987). By 70 d, most of the TEB have differentiated. The cancers induced during the sensitive period are first recognizable as intraductal proliferations within TEBs; these are usually initially multifocal. With growth they coalesce and evolve into predominantly papillary, cribriform or comedo adenocarcinomas. Alveolar buds either remain unchanged, give rise to hyperplastic alveolar nodules, or in some cases, form tubular adenomas. Cancers arising in irradiated rat mammary glands are also intraductal in origin as are those in irradiated mammary tumor virus-free BALB/c mice (Adams et al., 1987) and in chemical carcinogen treated mice (Medina and Warner, 1976). If DMBA is administered to 180 to 330 d old Sprague-Dawley rats, some of the adenomas and terminal ducts will give rise to well differentiated adenocarcinomas (Russo and Russo, 1987). In humans, breast cancers are classified as either ductal or lobular and the majority arise in the terminal duct section (Cardiff and Wellings, 1999; Wellings and Jensen 1973).

Human mammary glands from reduction mammoplasties and autopsies have been classified into three stages of differentiation on the basis of the morphology of their alveolar lobules (Russo and Russo, 1987). Type 1 or virginal lobules have approximately five to six alveoli each. These least differentiated lobules predominate in the glands of young and of nulliparous women. Type 2 lobules have approximately 40 to 50 alveoli each, and Type 3 have approximately 80 alveoli. Intermediate heterogeneous mammary glands composed only of Type 1 and Type 2 lobules or of ductal structures are found in some parous women, in women with histories of abortions and in those who have developed intraductal breast cancer in the contralateral breast. Type 3 alveolar lobules, the most differentiated class, predominate only in women who have carried a child to full-term.

Breast cancer in women, like that in rats and some strains of mice, arises primarily as intraductal hyperplasias that may be initially papillary and often progress to highly invasive cribriform and/or comedo carcinomas. A few arise within hyperplastic alveolar lobules at the periphery of the glands (Page and Anderson, 1987), and have a different prognosis. The risk of mammary cancer in women, as in rodents, is greatest in glands that have not progressed to full differentiation, that do not become populated primarily with Type 3 lobules characteristic of women who have carried children to full-term. Hence a woman who completes a full-term pregnancy soon after menarche lowers her breast cancer risk; pregnancy also markedly reduces susceptibility to carcinogen-induced mammary cancer in rats (Medina and Smith, 1999; Russo and Russo, 1987). The common types of spontaneous or radiation-related human breast carcinomas (Page and Anderson, 1987) resemble the most common intraductal type of rodent cancers induced by carcinogens (Huggins *et al.*, 1959; Kim *et al.*, 1960; Russo and Russo, 1987) or ionizing radiation (Clifton and Crowley, 1978; Ethier *et al.*, 1984). It is not clear whether the less frequent human peripheral alveolar lobule breast cancers are comparable to the adenocarcinomas that arise in tubular adenomas or terminal ducts of older rats treated with DMBA.

4.4.2 Hormones and Mammary Carcinogenesis

Mammary cancer risk is essentially abolished in women and rodents by ovariectomy soon after puberty. Conversely, about half of human breast cancers are responsive to estrogens, and some to estrogen and progestin. Rat mammary cancers can also be hormone responsive, particularly to hypophyseal prolactin (Clifton, 1979; Huggins *et al.*, 1959). It is difficult to separate estrogen and prolactin responsiveness in that estrogen stimulates prolactin release. In humans, the primary stimulant of growth of breast carcinomas is estrogen (Henderson *et al.*, 1988). It is clear, however, that estrogen and prolactin are synergistic in inducing mammary growth and differentiation. In the rat, prolactin alone stimulates mammary development in the absence of the ovaries. However, glucocorticoid is required as well as prolactin for the final steps in differentiation for milk secretion (Clifton and Furth, 1960). In addition, human chorionic gonadotropin stimulates functional differentiation of breast tissue to Type 3 lobules in pregnant women, increases the differentiation of TEB to alveolar buds and alveolar lobules in rats, and reduces the sensitivity of the rat mammary gland to cancer induction by DMBA (Russo and Russo, 1987).

Development of overt cancer from radiation or other carcinogen-initiated rat mammary glands is promoted by post carcinogen exposure to estrogens (Huggins *et al.*, 1959; Shellabarger, 1976) and/or prolactin (Clifton, 1979). The most effective hormonal milieu for such promotion is a combination of the mitogenic stimulation of elevated prolactin combined with blockage of terminal differentiation by glucocorticoid deficiency (Clifton *et al.*, 1985). Hormonal promotion not only increases mammary cancer incidence in irradiated rats, but decreases the cancer latency. Although the role of hypophyseal prolactin in human breast development and mammary cancer promotion is less clearly understood, lactation as well as early pregnancy reduces breast cancer risk (Russo and Russo, 1987). Conversely, prolongation of the nulliparous, fertile life span between menarchy and menopause with the consequent increase in cyclic exposure to hormonal stimulation of incompletely differentiated glands potentiates breast cancer in women (Clifton and Sridharan, 1975). The long latency of radiation-related breast cancers (Tokunaga *et al.*, 1984) also supports the conclusion that hormonal promotion plays an essential part in mammary cancer development in humans as well as in rodents.

4.4.3 *Cellular Origins of Mammary Cancer*

Russo and Russo (1987) described four morphologically distinct cells in rat mammary glands based on electron density, and cytological characteristics, three epithelial cell types (clear, intermediate and dark cells), and myoepithelial cells. Clear cells are infrequent and may be neuroendocrine cells. In TEBs, terminal ducts and alveolar buds, the dark cells are the most abundant and are hormone dependent. The intermediate cells are larger and do not appear to respond to hormones. With differentiation of TEB into alveolar buds, intermediate cells almost doubled in percentage with a concurrent decrease in the percentage of dark cells, and the cell-cycle times of both cell types were markedly increased. In chemical carcinogenesis in the rat there is reduction in dark cells and an increase in intermediate cells.

Dulbecco *et al.* (1983) used a battery of immunological reagents as markers to discern 10 epithelial cell phenotypes in rat mammary glands. They proposed a developmental relationship among these phenotypes beginning with totipotent stem cells and their dedicated progeny, prolumenal cells, and pro-myoepithelial cells in the end buds. Differentiation was postulated to progress down two pathways through two stages each of lumenal epithelial and of myoepithelial, and to terminate in alveolar lumenal, epithelial and myoepithelial cells. Smith and Medina (1988) described a widely distributed population of pale-staining relatively undifferentiated cells scattered in the epithelia of the ducts and TEB of mouse mammary glands from the late fetal stage onwards. On the basis of morphometric analyses, Chepko and Smith (1997) and Smith (1996) have proposed a schema of differentiation of both luminal epithelial cells and myoepithelium from the pale cells, and have proposed that they are mammary stem cells. Kordon and Smith (1998) derived CzechII mouse mammary cells with nuclear DNA "tagged" with clonally distinct mouse mammary tumor virus DNA sequences. The tagged cells, identifiable by southern blot analysis, gave rise to complete mammary glands when transplanted to pregnant mice. Furthermore, tagged cells from the primary transplants formed new glands of tagged cells when subtransplanted to pregnant recipients.

Clifton (1990) described subpopulations of clonogenic rat mammary epithelial cells characterized by their ability to form monoclonal functional multicellular structures when grafted into subcutaneous white fat pads of hormonally manipulated recipient rats. If the cells are irradiated (Clifton *et al.*, 1976) or exposed to N-methyl-N-nitrosourea (Zhang *et al.*, 1991) before grafting, and the recipients are appropriately manipulated to produce high levels of prolactin and low levels of glucocorticoids, mammary carcinomas arise within these clonal glandular structures. If the hormonal milieu includes high prolactin with normal or elevated glucocorticoid levels, the structures formed in grafts of untreated cells are predominantly milk-secreting alveoli. If there is high prolactin with glucocorticoid deficiency, ductal structures are formed (Kamiya *et al.*, 1991). Alveolar structures and ductal structures in primary grafts each contain cells capable of forming either alveolar or ductal structures on subtransplantation depending on the graft recipient's hormonal milieu (Kamiya *et al.*, 1998). Flow cytometric analyses showed the alveolar and ductal structures in the grafts to contain the same functional cell types as in homologous areas of glands *in situ* subjected to comparable hormonal milieus. Changes in the total numbers, concentrations, and radiation response characteristics of the clonogens during prepubertal, pubertal and postpubertal development and under different hormonal conditions in the mature animal have been investigated. The concentration of clonogens is much lower, and their radiation sensitivity is much greater in the glands of prepubertal animals than in glands of older rats; both increase markedly with sexual maturation (Shimada *et al.*, 1994). It is not yet known whether the high sensitivity of prepubertal clonogens to killing by radiation is accompanied by a high sensitivity to neoplastic initiation as in prepubertal and pubertal atomic-bomb survivors (Land *et al.*, 1993b; Tokunaga *et al.*, 1984). Finally, suspensions of rat mammary cells form multicellular duct-like and/or alveolus-like structures in culture in Matrigel (Kim *et al.*, 1993). Progress has been made in concentration of the clonogens by differential filtration, short-term culture, and immunofluorescence-activated cell analysis and sorting (Kim and Clifton, 1993).

Clifton and his coworkers have concluded that the mammary clonogens are stem cells capable of forming both secretory and nonsecretory lumenal epithelial and myoepithelial cells in the rat mammary gland. Further, these stem cells are the primary targets for radiation- and chemical carcinogen-induced initiation of cancer. It seems very likely that human mammary tissue contains similar stem cells that are the progenitors of breast cancers. If this is the case the rodent models are valid in this important aspect.

4.4.4 *Summary*

The mammary glands of laboratory rodents and humans bear many resemblances in morphogenesis and growth and development, including to some extent hormonal response to radiation and a relation of susceptibility of radiation-induced cancer to differentiation. A complete comparison of the molecular aspects of carcinogenesis in the breast of the human, rat and mouse has not yet been made. The importance of host factors such as hormones and the role of tissue factors is considerable and species-dependent differences must be taken into account. The introduction of transgenic rodents, in particular mice, and the current molecular technique being used in humans provide the tools for investigating mechanisms and therefore the opportunities and pitfalls in extrapolation.

4.5 Thyroid

4.5.1 *General Background*

Thyroid cancer accounts for ~3 % of all malignancies in children and 1 % or less in adults, and the female to male ratio is low. The mortality rate is low but the comparative aspects and the understanding of the physiology, the pathology, and the molecular changes in carinogenesis of this gland make it of interest in the development of methods of extrapolation.

Although physiology of the thyroid has been the subject of study in a wide variety of mammals including sheep, dogs and hogs

(Dumont *et al.*, 1980), probably for practical reasons, the most common models for carcinogenesis studies have been rodents. Because of the difficulty of surgical thyroidectomy of mice and the advantages of size for other procedures such as estimation of blood hormone levels, rats have been more commonly used than mice. Humans and rats are both more susceptible to radiogenic induction of thyroid cancer than mice (Clifton, 1986; Dumont *et al.*, 1980). In humans and rats, cancers arise within the follicular cell population and it has been postulated that they arise from clonogenic stem cells (Clifton, 1990; Dumont *et al.*, 1980). Although both species develop cancers with various morphologies, papillary carcinomas predominate in humans while follicular carcinomas are more common in rats. In humans, the much less frequent sporadic and hereditary medullary carcinomas arise from parafollicular or C cells (Hazard *et al.*, 1959).

The induction of thyroid cancer by external and internal radiation is well established (Ron, 1996; Shore, 1992). The dose response to low-LET radiation is considered linear but the effects of dose rate and the response to ^{131}I are not well defined (Holm *et al.*, 1988; Shore, 1992). Because of the marked gender- and age-dependency for this cancer, and the differences in the types of tumors in humans and rats, it is difficult to obtain matched sets of data suitable for extrapolation. In the report of a large study of the induction of tumors in rats after exposure to ^{131}I, the analysis was based on both papillary and follicular cancers. The relative frequency of the two types of tumors was not stated but the follicular tumors, as would be expected, predominated. The risk of induction of carcinoma by ^{131}I and external x rays was considered comparable and the dose response was approximately proportional to the square root of the dose (Lee *et al.*, 1982).

In the future, it may be possible to extrapolate the estimates of risk of thyroid cancer induced by internal emitters. However, the dependence of risk on age, gender, dose rate, protraction, and the lack of accepted risk estimates for humans make it a challenging task.

4.5.2 *Histogenesis of the Thyroid Gland and Thyroid Cancer*

The thyroid arises in humans, rodents and other mammals as a downgrowth, the thyroglossal duct, from the floor of the embryonic stomodeum. The bulbous end of the thyroglossal duct assumes the bilobed "H" shape around the ventrolateral surfaces of the trachea at the level of the thyroid cartilage, and the remainder of the duct and its attachment to the stomodeum involutes. Portions of the third and fourth pharyngeal pouches adhere to the thyroid surfaces to form the caudad and cephalad pairs of parathyroid glands; the parathyroids remain morphologically and functionally discrete.

The bulk of the thyroid tissue from the early fetal stage onward is composed of spherical follicles lined with a single layer of cuboidal to columnar epithelium and filled with proteinaceous colloid secreted by the follicular cells. The interstitial spaces between follicles contain the supporting stromal, vascular supply, and scattered calcitonin-secreting C cells of neural crest origin. The function of the C cells is controlled independently of the follicular cells; C cells are more resistant to radiation carcinogenesis.

4.5.3 *Thyroid Function and its Control*

In all vertebrates, the active hormonal products of the thyroid gland are triiodothyronine (T3) and thyroxin [tetraiodothyronine (T4)] (Clifton, 1986). The synthesis and secretion of T3 and T4 follow the same sequence. Iodide is actively absorbed from the serum against a gradient. Rapid utilization of the ions is catalyzed by a thyroid-specific peroxidase. The latter enzyme is also involved in the polymerization of tyrosine molecules to form chains of thyronines attached to the thyroglobulin (TGB) and stored in the follicular lumina. TGB synthesis and iodine organification occur at the apical follicular ends of the thyroid cells in the region of their characteristic microvilli. T3 and T4 secretion proceeds by endocytic uptake of TGB by the follicular cells, fusion of the TGB-containing vacuoles with hydrolytic endosomes, hydrolysis of TGB to release T3 and T4 in the vacuoles, transport of the vacuoles to the basal poles of the cells, and release of T3 and T4 into the blood stream. The ratio of T3 to T4 in euthyroid rat serum is approximately 1:40.

Steps in the process are stimulated in a time-dependent fashion by thyroid stimulating hormone (TSH) (Clifton, 1986; Dumont *et al.*, 1980), which is released from the anterior pituitary gland under hypothalamic control in response to a decrease in circulating levels of T3 and T4. Endocytosis of TGB begins within minutes after TSH occupies its receptors on the basal membranes of the thyroid cells, and T3 and T4 release soon follows. Within several tens of minutes, an increase in iodide uptake occurs. This process varies little from mammalian species to species, including mice, rats and humans. Finally, if TSH stimulation is sustained, thyroid cell proliferation is induced. Indeed, TSH is the primary hormonal promoter of thyroid cancer in irradiated individuals in those species that have been investigated, including humans and rats (Dumont *et al.*, 1980).

4.5.4 *Cellular Economy of the Thyroid Gland and the Origin of Cancer*

Thyroid cancer in humans and other animals is most likely clonal in origin (Clifton, 1990; Dumont *et al.*, 1980). After irradiation or other carcinogenic treatment, focal hyperplastic nodules develop. These may progress to adenomas and/or to malignant tumors. Indeed, the distinction between malignant and benign thyroid neoplasms is often difficult. If TSH secretion is suppressed after carcinogen exposure, progression to malignancy is slowed or abolished.

Experiments with quantitative rat thyroid cell transplantation support the hypothesis that the thyroid epithelium contains a subpopulation of clonogenic stem cells, and that, as in the mammary gland, these are the cancer progenitor cells (Clifton, 1990; Groch and Clifton, 1992a; 1992b). When monodispersed rat thyroid cells are grafted in thyroidectomized recipients (*i.e.*, into rats with elevated TSH levels) members of this cell subpopulation proliferate and differentiate to form multicellular functional follicles. If sufficient numbers of thyroid clonogens are grafted, hormonal feedback between the graft and the hypothalamic-pituitary system is reestablished, and normal T3 and T4 and TSH titers ensue (Domann *et al.*, 1990). The follicular units formed in the graft sites have been used as a terminal dilution assay endpoint to determine viable clonogen concentrations, and the radiation dose to stem-cell survival relationship and their post-irradiation intracellular repair capacities have been defined (Clifton, 1990). As in the rat mammary transplantation system, if the thyroid cells are irradiated before transplantation to rats with elevated TSH levels, radiogenic thyroid cancer develops within the clonal follicular structures in the graft sites (Mulcahy *et al.*, 1984).

Experimental goitrogenesis has been a useful tool for study of the thyroid cell subpopulation (Dumont *et al.*, 1980; Groch and Clifton, 1992a; 1992b). Any agent that interferes with the concentration of iodine by the follicular cells (*e.g.*, an iodine deficient diet, potassium perchlorate) or with the synthesis of T3 and T4 (*e.g.*, propylthiouracil, aminotriazole) brings about sustained hypersecretion of TSH in goitrogenesis. During the phase of rapid goitrous growth, the very high labeling index shows that most of the thyroid epithelium is involved (Dumont *et al.*, 1980). However, transplantation assays show that the total numbers of clonogenic stem cells do not increase during rapid aminotriazole-induced goitrous growth (Groch and Clifton, 1992a). With continued aminotriazole treatment, as the goitrous growth ceases and the goiter enters the plateau phase, the total stem-cell population per gland rapidly increases to establish a new population level. If aminotriazole treatment is withdrawn when goiters are large, the goiter regresses with massive cell loss. However, the assayable stem-cell population does not decrease during this time (Groch and Clifton, 1992b). Comparative transplantation studies were done of cells prepared from normal glands, from aminotriazole-induced goiters, and from perchlorate/low iodine diet-induced goiters, each containing different concentrations of clonogens. The clonogenic cells of the goiters were indistinguishable from those of normal glands in their capacities to respond to TSH with follicle formation and hormone secretion (Groch and Clifton, 1992a).

The molecular changes involved in radiation-induced thyroid cancer in humans and experimental animals are becoming increasingly better understood. Papillary thyroid cancer found in children exposed to high doses of radioactive iodine (Rabes and Klugbauer, 1998) and cancers in adults exposed to external radiation (Bounacer et al., 1997) have provided a stimulus and material to study the associated molecular alterations. Different forms of rearrangements and activation of the *RET* protooncogene as a result of radiation-induced chromosome damage have been detected (Rabes and Klugbauer, 1998). Tumors similar to human papillary thyroid cancer occur in transgenic mice with targeted expression of *RET/PTC1* (Jhiang et al., 1998). The evidence indicates that the mechanism of radiation-induced papillary thyroid cancers is similar in mice and humans.

4.5.5 *Summary*

With minor variations, the control of thyroid physiology and morphology, and of thyroid carcinogenesis, is essentially similar among mammalian species. Experimental evidence indicates that, like mammary cancer, thyroid cancer originates from clonogenic stem cells. The choice of experimental model can thus be dictated by convenience and expense. For these reasons and because a large amount of information about them is already available, the rat is the species of choice for most studies.

4.6 Skin

4.6.1 *Introduction*

The skin is a major organ and in the human adult accounts for ~3 % of the total body weight. In humans the skin covers a surface area of ~2 m². The structure is complex and consists of an epidermal layer derived from ectoderm and a dermis derived from mesenchyme. In the dermis there are structures such as hair follicles and sebaceous glands that are developed from infolding of the epidermis (Montagna, 1962).

The epidermis consists not only of epithelial, but also of Langerhan's cells, which are produced in the bone marrow and, in the case of humans, also melanocytes, which are thought to originate in the neural crest. In pigmented mice and rats, the melanocytes are in the dermis and the hair follicles, whereas in humans, melanocytes are in the epidermis and hair follicles. The dermis consists largely of collagen fibers and the cells are fibroblasts, mast cells, and cells associated with various types of vessels. Skin appendages, such as hair follicles, sebaceous, eccrine and apocrine glands, which are located in the dermis, are, however, epithelial in origin.

4.6.2 *Epidermal Cancers*

Obviously there are a number of different types of cells in the skin that are potential targets for cancer induction. The most common cancers in caucasians are epidermal and the most common of these are basal-cell carcinomas (BCC). About 90 % of the total skin cancers occur on the head and neck (Scotto *et al.*, 1983). Such tumors are usually listed as naturally occurring but, of course, are generally causally related to exposure to sunlight. In experimental animals in which the skin is protected by hair, "spontaneous" cancers of the epidermis are a rarity, but can be induced by ionizing radiation and other agents. The cell of origin of BCCs is presumed to be the basal cells in the epidermis and also cells in the hair follicles. BCCs can be induced in rat skin and in some strains of mice. In the case of the rat, the cell of origin is thought to be in the hair follicles and possibly in the sheath of the hair follicle (Burns and Albert, 1986). In patients with the basal cell nevus syndrome, BCCs occur and these patients are very susceptible to their induction by ionizing radiation.

The second most common type of carcinoma of the skin in humans is squamous-cell carcinoma (SCC). In humans, the cell of origin is thought to be the prickle cells of the epidermis. In rats, this type of skin cancer is induced by ionizing radiation, but with lower doses, less frequently than BCC; with high doses, it is the type of cancer that predominates (Burns and Albert, 1986). Based on the early reports of skin cancer in humans exposed to x rays (Frieben, 1992) where the doses were high, the cancers were SCCs and occurred after quite short latent periods. The skin cancers in humans associated with treatment of psoriasis by ultraviolet A and psoralen ultraviolet A are SCCs (Stern *et al.*, 1979). In mice, psoralen ultraviolet A also induces SCCs; however, studies have been carried out in hairless mice in which SCCs are the predominant type of induced skin cancer by ionizing radiation and UV radiation. Acanthomas and SCCs also predominate in most strains of mice exposed to chemical carcinogens (Holland and Fry, 1982).

4.6.3 *Melanoma*

Melanomas currently account for ~1 % of cancers in the United States and is one cancer for which the incidence is increasing.

There is as yet no unequivocal evidence that melanomas in humans are induced by ionizing radiation, but the possibility has been studied in the U.S. Department of Energy worker populations, especially at Lawrence Livermore National Laboratory (Austin and Reynolds, 1997), in the atomic-bomb survivors, and in tinea capitis patients exposed to x-ray treatments (Shore, 2001; Shore *et al.*, 2002).

Experimental study of radiation-induced malignant melanoma has been difficult because of the lack of an accepted experimental animal model. Various types of nevi, which are benign lesions that arise from melanocytes, can be induced in the skin of experimental animals, in particular, mice and guinea pigs. The induction of melanoma by UV radiation exposure of the opossum (*Monodelphis domestica*) has been reported (Kusewitt *et al.*, 1991).

4.6.4 *Tumors of the Dermis*

Radiation-induced fibrosarcomas have been reported after relatively high doses in experimental animals (Fry *et al.*, 1986; Hulse, 1962). The cell of origin of these tumors was considered to be the dermal fibroblast. Comparable fibrosarcomas are rare in humans, and soft tissue tumors only occur after large doses, making a comparison of the risks between humans and experimental animals difficult, if not impossible.

4.6.5 *Mechanisms of Epidermal Carcinogenesis*

Quantitative data for the induction of BCCs come from the study of tinea capitis patients treated with x rays (Shore *et al.*, 2002) and from the study of the atomic-bomb survivors (Ron *et al.*, 1998). Molecular changes have been studied in tinea capitis patients treated with x rays and who subsequently developed multiple BCCs in areas of the skin exposed to sunlight (Burns *et al.*, 2002). The findings suggest that the x rays deleted or inactivated one allele of the patched gene, thereby increasing the susceptibility for induction of a tumor by UV radiation by inactivation of the remaining allele of the patched gene. It is likely that if the initial x-ray-induced event is a deletion, it involves the nearby *XPA* gene, which is important in DNA repair. Mice deficient in the homologous patched gene (*Ptch+/-*) are susceptible to naturally-occurring and radiation-induced BCC induction (Aszterbaum *et al.*, 1999; Mancuso *et al.*, 2004). It is clear that the heterozygosity of this gene has similar effects in humans and mice.

4.6.6 *Importance of Interactions*

Human skin is relatively resistant to the induction of epidermal cancers by ionizing radiation alone, but is quite susceptible when exposure to ionizing radiation is followed by protracted exposure to UV radiation (Shore, 1990). The importance of the interaction of ionizing and UV radiation in the risk of skin cancers in humans poses a further complexity to extrapolation of risk estimates of cancer induction in this organ from animals to humans.

4.6.7 *Summary*

Experimental model systems exist for the main types of epidermal cancers that occur in humans and, provided the data for humans are not confounded by interactions, it may be possible to test methods of extrapolation.

4.7 Gastrointestinal Tract

4.7.1 *Introduction*

Cancers of the esophagus, stomach, small intestine, colon, and rectum account for ~15 to 18 % of the total cancer mortality in the Western world but ~30 % in Japan. Cancers of the GI tract occur less frequently in unexposed and exposed experimental animals. Characteristically, high doses are required to induce cancers of the GI tract in mice and rats.

4.7.2 *Stomach*

In humans there are two main types: (1) relatively well differentiated carcinomas that show intestinal metaplasia; ~80 % of the tumors in Japan are of this type, and (2) poorly differentiated scirrhous carcinomas, which are more common in the Western world than in Japan.

In mice and rats, the histological structure of the stomach differs from that in humans as it is divided into a forestomach consisting of a squamous epithelium and a glandular portion quite similar in structure to that in humans. Spontaneous tumors of the stomach are very rare (Hare and Stewart, 1956). Adenocarcinomas of the glandular portion and SCCs of the forestomach have been induced in mice by radiation (Cosgrove *et al.*, 1968; Nowell and Cole, 1959). In the rat, most of the reports of carcinogenesis of the stomach relate to chemicals. Adenocarcinomas in the rat are more differentiated than those in humans and metastasize less frequently. The target cells for adenocarcinomas are probably the same in humans, rats and mice.

4.7.3 *Small Intestine*

Tumors of the small intestine are remarkable for their rarity in humans and experimental animals although it consists of a very large-cell population maintained by a high proliferation rate of cells in the crypts of Lieberkuhn. The cell turnover rate is somewhat lower in humans than in rodents but the number of stem cells in the small intestine is large.

In humans only 1 % of all tumors of the GI tract involve the small intestine; of those, almost 50 % are benign. In mice, focal hyperplasia, such as duodenal plaques or polyps, are relatively frequent and their incidence is influenced by genetic background and diet (Hare and Stewart, 1956). The very rare carcinomas in mice arise through precancerous changes in the crypt epithelium, and adenocarcinomas have been induced by exposure to neutrons (Nowell and Cole, 1959; Nowell *et al.*, 1956).

In rats the type of epithelial tumor is related to the site of origin. For example, adenometous divertula arise in the deodenun. Adenocarcinomas of low-grade malignancy have been induced with large doses of radiation (>10 Gy) (Bond *et al.*, 1952; Brecher *et al.*, 1953; Marks and Sullivan, 1960; Osborne *et al.*, 1963).

4.7.4 *Colorectal Tumors*

The incidence of adenocarcinoma of the large intestine in the Western world is about 1 in 200 and in the United States accounts for ~15 % of all cancers. In contrast, tumors of the large bowel are rare in rodents. Most of the tumors in humans and rodents start as polyploid lesions and progression from adenoma to carcinoma seems to be the rule. There are a number of different types of adenocarcinomas in humans such as mucinous, signet-ring cell, squamous cell and adenosquamous. Morphologically and in their behavior, intestinal tumors induced in rats resemble the corresponding tumors in humans (Denman *et al.*, 1978).

A dominant mutation that predisposes to multiple intestinal neoplasia (*Apc*) has been found in the mouse (Moser *et al.*, 1990). The tumors start as adenomas and progress to carcinomas. The gene responsible is the homolog of the adenomatous polyposis coli (*APC*) gene in humans. A further gene *Pla2q2a* (Dietrich *et al.*, 1993), which modifies the expression of the tumors, has also been identified. It appears that some of the tumors in humans and mice have a similar interaction of the *APC* gene in humans and the homozygous *Apc* in mice (Degg *et al.*, 2003; Levy *et al.*, 1994).

The genetic predisposition to both naturally occurring and radiation-induced tumors in the small and large intestine in humans and mice is complex (Ruivenkamp *et al.*, 2003) because of the number of loci that influence susceptibility.

4.7.5 *Summary*

Induction of GI cancers in experimental animals requires relatively high doses, and the estimates of risk are not precise. In the atomic-bomb survivors, cancers of the stomach, large intestine, and rectum contribute a significant fraction of the total radiation risk (ICRP, 1991). The apparent difference between the Japanese population and experimental animal populations suggests the influence of factors peculiar to humans or that particular population since the cells of origin are unlikely to be significantly different. It may be difficult to identify a human population in which there is no factor confounding the radiation risk. The mice models that have been developed appear to be excellent for the study of carcinogenesis, but not necessarily for deriving methods of extrapolation.

4.8 Bone

4.8.1 *Humans*

Tumors that are classified as bone tumors may arise from cartilage, fibrous tissue, or osteoid matrices. Cartilage is the most common site of origin. Tumors may be either benign or malignant, but malignant tumors arising in the skeletal system are rare, accounting for a mere 0.2 % of all primary cancers. Osteosarcomas and

Ewing's sarcoma are the most common forms of malignant bone tumors, and they occur mainly in childhood and adolescence (Dahlin, 1978; Lichtenstein, 1977). The induction of bone cancer in humans has been associated with exposure to external radiation, and to internally-deposited radium (Rowland, 1994; Rowland et al., 1983); there is very little information about the effects of viruses or chemical agents.

In investigations of the histogenesis of bone tumors in humans, it was reported that in naturally occurring tumors 55 % were osteosarcomas, 35 % were chondrosarcomas, and 10 % were classified as malignant fibrous histiocytoma/fibrosarcoma. The corresponding percentage values for the tumors arising after external irradiation were 63, 33 and 4, and after exposure to ^{224}Ra, they were 53, 33 and 14, respectively (Gossner, 1999). A comparable documentation for the tumors in experimental animals should be possible.

The two most common histological types of tumors in patients administered ^{224}Ra are osteosarcomas, which produce bone, and fibrous-histiocytic sarcomas, which do not produce bone. A similar spectrum of bone cancers occurs in populations of humans that were exposed to other isotopes of radium. The important point is that more than one type of target cell can be involved in radiation-induced bone tumors.

4.8.2 *Mice*

A dual system has been used to classify neoplasms of bone in the mouse. The first is based on the most probable progenitor cell, the second is based on the behavior or malignancy of the tumor. A problem is the occurrence of several neoplastic tissue types in a single tumor. For practical purposes, bone tumors in the mouse have been classified according to their resemblance to those in humans and dogs. The benign tumors arising from osseus tissue are osteoma, ossifying fibroma, and osteofibroma (Nilsson and Stanton, 1994).

It is often assumed that bone tumors arise only from dividing cells but it is not known whether cells at various stages of the osteogenic cell lineage are equally susceptible or whether the histological characteristics of the tumors are dependent on the stage of differentiation of the irradiated cell or the tissue environment.

The most common malignant bone tumor in mice is the osteosarcoma. The malignant tumor arising in cartilage, the chondrosarcoma, is extremely rare in mice. Osteosarcomas are classified on the basis of whether or not bone formation is predominant, as being either osteoblastic or fibroblastic, respectively. However, the tumors are usually heterogenous in structure and consist of collagenous, cartilaginous and osseus elements.

Tumors of bone in mice can be induced by external radiation, internally-deposited radionuclides, chemical agents including estrogen, and viruses. In mice, induction by internally-deposited radionuclides (Gossner et al., 1976; Vaughan, 1973) and the role of viruses have been studied extensively (Erfle et al., 1986; Finkel et al., 1973; Lloyd et al., 1975; Stanton, 1979). The distribution and histogenesis of bone tumors induced by viruses differ from those in radiation-induced tumors and appear to be virus specific. The retroviruses that cause bone tumors are related to the murine leukemia-sarcoma virus complex (Kelloff et al., 1969) and the associated oncogene is v-*Fos* (Curran and Verma, 1984). Murine bone tumors appear to be influenced greatly by certain viruses described by Finkel et al. (1973), but this does not appear to be the case in humans. There have been no reports of an involvement in murine bone tumors of *Rb*, *Trp53* and other genes implicated in the induction of such tumors in humans.

Osteosarcoma is rare in mice but very high incidences can be induced by alpha-particle emitting bone-seeking radionuclides. The naturally occurring osteosarcomas in humans and mice are histopathologically indistinguishable. Although there are a large number of genes involved in osteosarcomagenesis (Miller et al., 1996; Nathrath et al., 2002; Rosemann et al., 2002; Sturm et al., 1990), the genetic alterations appear very similar in the radiation-induced tumors in mice and the naturally occurring osteosarcomas in humans (Rosemann et al., 2002).

4.8.3 *Rats*

The experimental induction of bone tumors in rats has been studied mainly in relation to the effects of radionuclides (Evans et al., 1944), but they have also been induced by viruses (Kirsten et al., 1962) and by chemical carcinogens (Kalman, 1970). For review of bone tumors in the rat see Litvinov and Soloviev (1987). Bones in the rat, unlike those in humans, continue to grow throughout the animal's life. Naturally occurring or spontaneous tumors are extremely rare, but bone cancer is quite susceptible to induction by relatively large doses of external radiation, internally-deposited radionuclides, and chemical carcinogens. Although the induced tumors show similar proportions of fibrous and osteogenic tissue as in mice, they are usually classified in rats simply as osteosarcomas or osteogenic sarcomas.

4.8.4 Dogs

The dog has been the experimental animal of choice for the study of the induction of bone tumors by internally-deposited radionuclides. The beagle has been used extensively in studies of the induction of bone tumors by radium and plutonium (Dougherty et al., 1962), and the data from these studies are the main basis of risk estimates of bone tumors induced by these internally-deposited radionuclides in humans. As in the other species discussed, the majority of the induced tumors are osteosarcomas, but other types such as chondrosarcomas and myelomas have been reported (Brodey et al., 1963; Lloyd et al., 1991). The locations of naturally occurring and putative ^{239}Pu-induced osteosarcomas in the skeletons of humans have been compared with those in the beagle dog. The distribution of the plutonium-induced tumors was similar in both species, and the most frequent site was the spine. The distribution of the naturally occurring osteosarcomas is quite different, the tumors being in the peripheral skeleton in humans and beagle dogs (Miller et al., 2003). Bone tumors do occur naturally in dogs, and the Saint Bernard appears to be particularly susceptible. However, the susceptibility for induction of bone tumors by radiation is not correspondingly greater, than say, in the beagle. Furthermore the sites of radiation-induced bone tumors in the Saint Bernard dog are not the same as the sites of the naturally occurring tumors (Taylor et al., 1997).

4.8.5 Summary

The data obtained from studies of the induction of bone tumors by plutonium and radium have been used to extrapolate the risk of bone cancer determined from experimental data for dogs (Muggenburg et al., 1983) to the risk in humans using the relative toxicity ratio method (Evans et al., 1944; Mays et al., 1986a), a Bayesian approach (DuMouchel and Groer, 1989), or a scaling method (Raabe, 1989). These methods and the experimental methods are discussed later in Section 6.

The original maximum permissible body burden of 1.5 kBq (0.04 µCi) of ^{239}Pu was derived from studies of the comparative carcinogenicity of ^{239}Pu and ^{226}Ra in rabbits, rats and mice (Brues, 1951). More recently, beagle dogs have been the animal of choice. The target cells for osteosarcoma are the same type in the various species. Although the distribution of tumors is somewhat different between beagles and humans, it is clear that the location where the dose is delivered determines the site where tumors occur. Not enough is known about the mechanism of carcinogenesis in bones of dogs and humans to discern whether any differences in the mechanisms pose a stumbling block to extrapolation.

The fact that the cells at risk are likely to be the same in both species, the molecular changes in osteosarcomas are remarkably similar in humans and mice and, in general, the behavior of radiation-induced tumors appears to be similar suggests that extrapolation of the risk across these two species may be possible.

Dose-response relationships of cancer induction are based mainly on the occurrence of cancer at specific sites such as breast or skin, without any designation of the type of cell from which the cancer originated. This is unfortunately the case for human and experimental animal data. When methods of extrapolation of risk estimates are applied, it is important to know the basis of the tumor classifications that have been used for the derivation of the dose-response relationship. For example, the term "lung cancer" is not adequate because the responses and mechanisms of induction are likely to be different for alveologenic carcinomas in mice and small-cell carcinomas in humans. Furthermore, the crude classification, "lung cancer," does not reflect the fact that alveolar carcinoma, while by far the most common type in mice, is not the dominant type in humans, and that small-cell carcinoma, while common in humans, is not found in mice.

The disparity and similarity between humans and experimental animals in the types of cells that are the targets for the induction of cancers by radiation at specific sites vary. As indicated above, any attempt to extrapolate the risk of induction of "lung cancer" by radiation from mice to humans must take into account that in mice the spectrum of the types of lung cancers differs from that in humans. In other tissues, as the individual subsections in this Section indicate, the probability that the target cells are similar in experimental animals and humans is high. For example, in bone tumors, cells in the ostogenic lineage are almost certainly the cells at risk for the induction of osteosarcoma in humans, dogs and rodents. Many of the tumors induced by radiation have a different histogenesis than that of osteosarcomas and, while the target cell may not be the same as for osteosarcoma, the target cell may well be the same in the different species. It is also possible that pluripotent stem cells are the target for both types of tumor. A comparable case may be made for lung tumors. Markers that distinguish small-cell carcinomas from nonsmall-cell tumors have been found. However, there is considerable heterogeniety in the expression of the markers and overlap of markers among tumors of different cell types (see review by Minna *et al.*, 1989). Here again there is the possibility of a common cell of origin, presumably a pluripotent stem cell with the potential for multiple paths of differentiation. If this turns out to be the case the apparent differences of cell of origin may not be a barrier to extrapolation of risk estimates derived from data for different organs, for example, those based on the incidence of lung tumors rather than adenocarcinomas of the lung.

Thus, in a number of tissues the similarity in the target cell and even the mechanism of the induction of the initial events may be sufficient to encourage the search for methods of extrapolation. In some tissues, for example breast, the target cell and initiation may be similar in more than one species, but the processes involved in the expression of the initial event may differ. Therefore, in the case of such tissues, the method of extrapolation must take into account and overcome these differences in the factors influencing expression.

The hope of developing methods of extrapolating risk estimates across species has been reinforced by the continuing evidence of the conservation of the genes involved in the control of cell proliferation, DNA repair, and other aspects salient to the induction of cancer. Furthermore, the lesions, whether they be translocations, deletions or point mutations, appear to be similar in mice and humans, as do the DNA sequences involved in specific cancers.

The information about tissues in different species in this Section is encouraging and in those tissues where differences are major it should serve as a guide to other approaches. The degree of detail about the various tissues and the mechanisms of carcinogenesis vary markedly in the various subsections of this Section. For example, the section on the hematopoietic system is much more detailed than the sections on solid tissues. The reasons for this include the fact that: (1) leukemia is a characteristic radiation-induced cancer, and (2) there are a number of different types of leukemias and lymphomas. Furthermore, the detailed discussion about leukemias serves to illustrate the complexity of the problem, especially with regard to the mechanisms of action in different species. The GI tract has been discussed because tumors of the GI tract dominate the cancer risk estimates for the atomic-bomb survivors. In contrast, experimental animals that have been studied appear to be very resistant to the induction of cancer of the GI tract by radiation and also to have a low natural incidence. In the intestinal tumors in the *min* mouse, the *Apc* gene is involved, as is the case in the tumors of the large intestine of humans.

The relationship of natural incidence of tumors to the susceptibility for induction by radiation is an important question. While it appears that a positive relationship does not exist for all types of cancer, it does for some. A corollary of this question is whether such a relationship would favor the approach of extrapolating relative risks. It was found that risks for some solid cancers could be extrapolated across strains of mice but not the risks of lymphomas (Storer *et al.*, 1988).

The risk estimates that have been developed for radiation protection are based on a general or working population that spans a considerable range of ages, whereas animal experiments on the induction of cancer have generally been carried out with exposure to radiation at one age, or protracted over a specific age range starting with all the animals at the same age. Not only must the radiation exposure patterns be kept in mind, but in the case of rodents, most of the pertinent data come from experiments on one inbred strain or hybrids of two inbred strains, each with their individual spectrum of susceptibility and resistance genes. Neither of these choices appears ideal for extrapolation purposes unless, as posed above, relative risk can be shown to be an appropriate approach.

5. Radiation Effects at the Molecular and Cellular Levels

5.1 Introduction

To predict the adverse health outcomes of exposure to ionizing radiations for humans, it is frequently necessary to utilize data obtained from laboratory animals and, by using a series of correlation factors, extrapolate to the anticipated response(s) in humans. This type of approach is necessary, of course, because there is relatively little information on the induction of somatic and genetic effects by ionizing radiations in humans. A frequently proposed method of estimation involves the direct extrapolation from animal tumor or birth defect data to humans. This extrapolation assumes a common mechanism of formation of the endpoint in the different species, which includes cell-type of origin for tumors and potentially confounding host factors, for example. While this would appear to be the most effective approach available, it is clearly an equivocal one, largely because of the assumptions made and the necessity of having complete and appropriate animal data to form the basis for extrapolation. It would seem to be of particular importance at this time to consider if or in which ways extrapolation of genetic and somatic risk to humans from animal data could be markedly improved by incorporating an understanding of the mechanism of formation of endpoint (tumor or birth defect) into the approach. Since this can only be partially feasible because of the current incomplete nature of our knowledge, it is sensible to consider the use of surrogate endpoints for cancer, namely chromosomal alterations and mutations. This is reasonable since specific genetic alterations are clearly involved in the production of tumors and birth defects. The following two sections will address the mechanism of induction of chromosome aberrations (structural and numerical) and point mutations by ionizing radiations, how such information might allow for comparison among various species, including humans, and what factors could influence sensitivity. In addition, the involvement of specific gene alterations in neoplastic cell transformation is discussed. This approach represents the possibility for a direct means of estimating human tumor responses from exposure to ionizing radiations since information on the induction of the specific gene alteration might be used to predict tumor responses. The emphasis will be on comparative responses among species, known or predicted.

5.2 Effects of Ionizing Radiations at the Molecular Level

5.2.1 *DNA Damage*

Despite many years of study, the specific nature of damages to DNA induced by ionizing radiations of different qualities is only now beginning to be determined. In a somewhat simplistic fashion, it is most common for DNA damages induced by ionizing radiations to be classified as single-strand breaks (ssb), double-strand breaks (dsb), base damages (bd), DNA-DNA and DNA-protein crosslinks, and clustered damage. The clustered damage as proposed by Ward (1994) and Goodhead (1994), for example, is presumed to include all the various types of DNA damage within a limited stretch of base pairs (Sutherland *et al.*, 2000; 2002). There has been a wide range of quantitative studies for ssb and dsb that has assessed the yields of breaks under a variety of conditions, and for a number of radiation qualities for a broad range of species (Roots *et al.*, 1990; Teoule, 1987; Whitaker *et al.*, 1991). While this does have some merit, it is to be noted that comparison among cell types, species and radiation qualities has to rely upon the general assumption that all strand breaks are created equal, whereas this is almost certainly not the case, at least as far as probability of conversion of different types of dsb, for example, into chromosome alterations and/or point mutations. (Pinto *et al.*, 2002; Stenerlow *et al.*, 2002). Certainly this probability would be expected to be quite different for strand breaks induced by radiations of different quality. Data on mutational spectra at the DNA sequence level might be able to address this problem, but to date no specific pattern of deletions, rearrangements or point mutations can allow for any particular conclusion about the nature of the dsb that might have been responsible for the formation of the mutation. It is probably reasonable to assume that the spectrum of induced strand breaks among species and cell types is quite similar (Rothkamm and Lobrich, 2002). Further, it is also assumed that the frequency of ssb and dsb will be proportional to the DNA content of a cell. It has been broadly shown that this is the case for all types of radiation studied (Frankenberg-Schwager, 1990). However, the ratio of ssb to dsb, and the absolute number of strand breaks per unit amount of radiation is quite different for radiations of different LET, with generally a higher proportion of dsb at higher LETs (Goodhead, 1994). This would seem to be consistent across a range of species.

Information on the types and frequencies of bd under various radiation conditions and for different radiation qualities is still quite limited, despite a recent increase in activity. There is a very large number of different types of bd induced by ionizing radiations (for reviews see Teoule, 1987; Wallace, 1988), but the majority of the qualitative and quantitative data are for thymine glycol, 8-hydroxydeoxyguanine and urea, largely because the appropriate detection methods are available. Suffice it to say that from theoretical considerations, the amounts of bd would be expected to be in proportion to DNA content of a cell type for radiations of different quality. Whether or not the spectrum of changes would be the same for different radiations is unknown at this time, although one might expect some differences.

The relationship of the proportion of strand breaks to bd has been estimated for low- and high-LET radiations, and suggests that at low-LET the ratio of ssb:dsb:bd is about 25:1:50, respectively, but at high-LET the ratio of ssb:dsb is closer to 1:1, respectively, with little or no information being available for bd (Goodhead, 1994; Ward, 1988). These sorts of data provide a starting point for considering relative sensitivities of different cell types and similar cell types in different species to the induction of genetic alterations that would be predicted to be involved in the production of a tumor. However, it is less likely that induction of specific types of DNA damage will influence sensitivity, but rather that the sensitivity will be affected by the manner in which these damages are handled by the cell; *i.e.*, they can be repaired efficiently, they can be repaired incorrectly, at some frequency, and they may not be repaired, at least not in a timely manner. This leads quite naturally to a consideration of the repair of radiation-induced DNA damages.

5.2.2 *Repair of DNA Damage*

In the past 5 to 10 y there has been a considerable increase in our knowledge of the molecular aspects of DNA repair. This includes a very complete understanding of the nucleotide excision repair pathway for UV-induced DNA damage (Sancar, 1995; Wood, 1995; Wood *et al.*, 2001), of the repair of specific chemical adducts by, for example, methyl guanine methyl transferase

(Grombacher *et al.*, 1996; Harris *et al.*, 1991; Pegg, 2000; Tano *et al.*, 1990) and oxidative DNA damage (Demple and Harrison, 1994; Fortini *et al.*, 2003). More recently information has been gathered on the repair of dsb and the relationship of one of these, nonhomologous end-joining (NHEJ), to recombination of immunoglobulin genes (Jackson, 2002; Jeggo *et al.*, 1995; Kanaar *et al.*, 1998). What is apparent is that for ionizing and nonionizing radiations the repair pathways are complex and involve multiple enzymes. This is a consequence of damage recognition and repair pathways being exquisitely accurate, and perhaps because DNA repair pathways were derived from normal cellular housekeeping functions. Much of the data on the repair of ionizing radiation DNA damage remains phenomenological, although it does allow for a consideration of the relative effectiveness of the repair process(es) in different cell types and species.

5.2.2.1 *Single-Strand Breaks.* In general, the repair of single-strand breaks (ssb) is by simple religation of the broken ends with modification of the broken ends being necessary to produce a 3-OH and a 5-phospho group. Loss of a nucleotide during this broken end modification process presents no great problem to the cell because a presumably undamaged base is present on the complementary strand allowing "fill in" repair. This process of repair is predicted to be largely error-free and quite rapid. In fact, the majority of ssb are repaired in a matter of a very few minutes even when doses of tens of gray have been delivered (Van der Schans *et al.*, 1983). There is little or no evidence to suggest that the misrepair of ssb is involved in the production of mutations, and since their repair is rather rapid, there is little expectation that they would be involved to any great extent with mutation induction by errors of DNA replication. It has been shown by Natarajan and Obe (1978) that if x-ray-induced ssb are converted into dsb by means of an introduced *Neurospora* single-strand endonuclease, then an increase in chromosome aberrations results. However, this observation shows that dsb are of importance in the formation of chromosome alterations, rather than implicating the conversion of ssb into dsb as a normal cellular phenomenon. Thus, for a consideration of a mechanistic approach for extrapolating across species for endpoints that are related to disease processes, it would seem to be reasonable to exclude ssb induction.

5.2.2.2 *Double-Strand Breaks.* More information is available on the repair of double-strand breaks (dsb), to a large extent because such lesions can be introduced in a defined manner by restriction endonucleases and because many of the genes involved in dsb repair have been identified (Jackson, 2002). Much of the information still remains pertinent only for high exposures (several gray) since the frequencies of dsb are low even at these high exposures. It is important to note also that radiation-induced dsb would be much more variable with regard to cut-end structure and DNA sequence location than those produced by restriction endonucleases. Thus, any correlations between repair of dsb produced by the two methods should be made with some caution.

There are two major pathways involved in the repair of DNA dsb, homologous recombination and NHEJ (for a review, Jackson, 2002). Homologous recombination is the major repair pathway in yeast, whereas NHEJ is the major one in mammalian cells.

5.2.2.2.1 *Nonhomologous end-joining.* A dsb can be repaired by ligation of the cut ends, usually with some modification of these. It is unlikely that radiation-induced dsb will be equivalent to the blunt-end type of lesion produced by some restriction endonucleases but rather more similar to the cohesive or overlapping-end type, bearing in mind that there is unlikely to be a specific DNA sequence among radiation-induced dsb in contrast to restriction enzyme-induced dsb. This consideration will clearly be of importance when considering the process of misrepair or misjoining. The various modes of religation of restriction enzyme-induced dsb have been discussed by Goedecke *et al.* (1994) and Pfeiffer and Vielmetter (1988). Cohesive ends will be the major kind of dsb induced by radiation and those are repaired by the nonhomologous end-joining (NHEJ) process. It should further be noted that a dsb can be formed when two ssb occur within a region of five to seven base pairs (Van der Schans, 1969).

It is of significance to the integrity of the genome that repair of dsb be accurate and restore the original DNA molecule. In terms of genetic alterations, it is important to consider the consequences of inaccurate repair. For the further consideration of the relative sensitivities of cells from different species to mutation induction, for example, it is necessary to determine the relative probabilities of misrejoining during religation of dsb for the different species (for general review see Rothkamm and Lobrich, 2002).

The probability of misrepair during NHEJ of dsb would be expected to be influenced by the frequency of dsb, the length of time that dsb remain unrepaired, the cellular distribution of the dsb, the fidelity of the ligation process, the relative proportion of the repair of dsb that occurs *via* this process, and, hence also, the nature of induced dsb (related largely to radiation quality).

Which of these factors would be expected to influence interspecies variations in sensitivity to mutation induction? It has already been discussed above that the frequencies of dsb, their nature, and cellular distribution would not be predicted to be very different among species, except that the overall frequency will be dependent upon DNA content of the cell. The fidelity of the ligation process could vary among species, but since whether or not a particular dsb is correctly ligated or cross-ligation occurs with a neighboring dsb probably depends most significantly upon dsb distribution and thus proximity of dsb, then there would not be expected to be much difference among species. It should be noted that in order to incorrectly ligate radiation-induced dsb (since they have cohesive-ends that would generally be nonhomologous) one or both cut ends would need to be excised prior to ligation. This will frequently lead to the deletion of a few bases, at the site of formation of a chromosomal rearrangement (translocation or deletion, for example).

Failure to repair a dsb by NHEJ could also lead to a mutation, most likely a deletion (but possibly a point mutation as the consequence of a replication error at the site of the unrepaired dsb). While there is little or no information that would indicate a difference in completeness of ligation among species, there are some data that indicate that such differences can occur within a species. Cells from individuals with the radiation-sensitive syndrome ataxia-telangiectasia (A-T) are efficient at rejoining dsb but they do so with a higher frequency of misrepair than do normal cell lines (Cox et al., 1986). It can be assumed from the experimental design that the introduced dsb was either incorrectly rejoined or nonrejoined. Whether this results directly in the radiosensitive phenotype is, of course, conjecture, although information on the function of the gene mutated in A-T individuals suggests that this would be an indirect correlation (Rotman and Shiloh, 1998).

On the assumption that the chromatin of a cell is not static, it would be expected that the longer a dsb remains unrepaired, the greater the chance that it will be in contact with an adjacent DNA molecule and subsequently produce a misjoining event. While the analysis of the repair of dsb has not distinguished among the various modes of repair, there is no indication that overall repair of dsb differs significantly among species. Again of note, there are a number of reports of two types of dsb, one that is rapidly repaired and one that is slowly repaired (Fox and McNally, 1988; Metzger and Iliakis, 1991). From the above arguments it would be predicted that the slowly repaired dsb would be more likely to be involved in misjoining events leading to genomic alterations. The present data are insufficient for determining whether or not the proportions of these different "types" of dsb vary among species. It could be of interest to determine this as part of an approach to predicting interspecific variations in sensitivity to mutation induction.

It is proposed by Goodhead (1994) and Sutherland *et al.* (2000; 2002) that clustered damage that might typically include multiple dsb along with other damages is resistant to repair, and perhaps, especially when induced by high-LET radiations, irreparable.

5.2.2.2.2 *Recombination repair.* It has also been established that dsb can be repaired by a recombination process (Szostak *et al.*, 1983). This has been most clearly demonstrated for the process of mitotic recombination or homologous recombination where an initiating event is an endonuclease produced dsb. Although it should be noted that this process is not for the repair of dsb *per se*, but rather they are repaired as a necessary component of the recombination. The evidence for recombination repair of radiation-induced dsb in mammalian cells is becoming more clearly defined (Thompson and Schild, 2002), and extrapolation from data in prokaryotes and yeast (Friedberg *et al.*, 1995) can enhance our understanding of this mechanism. Johnson *et al.* (1999) data have indicated that a specific gene (*XRCC2*) is involved in recombination repair of radiation-induced dsb, and the essential role of the Nijmegen breakage syndrome gene (*Nbs1*) has recently been described (Tauchi *et al.*, 2002).

It is apparent that the relative efficiency of a recombinational repair process among cells from different species will depend upon the relative effectiveness of a number of steps in the process; strand invasions, exonuclease activity, and ligation, in particular. Further, it should be emphasized that recombinational repair can involve intrachromosomal or interchromosomal (homologous sequences on nonhomologous chromosomes) recombination. The former can lead to no mutations or a variety of point mutations, deletions and rearrangements depending upon the fidelity of the process. The latter (interchromosomal) can result in translocations, deletions and point mutations again as a consequence of the process itself, where nonhomologous chromosomes are involved, and also as a consequence of the fidelity of the process. There is insufficient information for it to be determined whether or not there are interspecific differences in the extent, fidelity or nature of the recombinational repair process. As a very indirect piece of information, there are differences reported in excision repair efficiency for UV-induced DNA damage among species (Hart and Setlow, 1975), and since a number of the same or similar enzymes are involved in the two modes of repair it could be speculated that there will also be large differences in recombination repair efficiency and/or fidelity among species.

It would, as suggested above, be important to determine whether a dsb will be repaired by NHEJ or a recombinational event. Although this is not clearly defined at the moment, there is some information that addresses this matter. For example, it has been shown that dsb in transcriptionally active genes are preferentially repaired in human and rodent cells, and that this repair is *via* a recombination process (Frankenberg-Schwager *et al.*, 1994). The constraints that chromatin structure might impose on recombination could determine the pathway chosen for repair. Lopez and Coppey (1987) showed that the end structure at the break site could be a determining factor as to which method of repair a cell will utilize. For example, dephosphorylation of the dsb ends prevents the ligation pathway but does not affect the recombination pathway; blunt or 5'-protruding ends can be repaired by recombination, whereas 3'-overhanging ends cannot. This could be a consequence of the polarity of the exonuclease and recombinase activities associated with the repair process. Short stretches of perfect homology rather than long stretches of partial homology were found to govern the efficiency of recombination. Wahls *et al.* (1990) found that when a dsb was introduced into hypervariable minisatellite sequences, recombination was stimulated over and above that caused by dsb in other sites in the plasmid used. Efficient recombination in mammalian cells can be maintained with a homologous stretch of DNA as short as 165 to 320 base pairs and with a low level of recombination measurable for 27 base pairs of homology (Lopez *et al.*, 1992). This would allow for the recombination to occur between nonhomologous chromosomes that contain relatively short sequences of homologous DNA, such as minisatellites or other repetitive DNA segments. The consequence would be chromosomal rearrangements and/or deletions. Determining the balance between NHEJ and recombination repair among species is of importance when considering the mechanistic basis for radiosensitivity.

5.2.2.3 *Base Damage Repair.* It is generally agreed that the major pathways by which DNA base damages (bd) are repaired are: (1) *via* a glycosylase leading to an apurinic or apyrimidinic site that can be filled by an appropriate base, or can be removed by an excision repair pathway or (2) removal of the damaged base by an excision repair pathway (Demple and Harrison, 1994). In this regard, it is interesting to note that Satoh *et al.* (1993) showed that some specific bd, produced by gamma radiation and H_2O_2, are inefficiently removed by XPA cells that are defective in the nucleotide excision repair pathway. This suggests that perhaps more than one excision repair pathway is involved in the repair of DNA bd; this suggestion is not surprising given the variety of types that are induced.

Because of the difficulty of assessing the repair of the myriad of bds induced by ionizing radiations little information on their repair has been forthcoming. The development of antibodies to DNA bd has provided some information on the kinetics of repair and consequences of replication on a template containing a specific bd (Demple and Harrison, 1994; Hubbard et al., 1989; Leadon and Hanawalt, 1983; Rothkamm and Lobrich, 2003). A report by Le et al. (1998), described an ultrasensitive method for measuring bd that used immunochemical recognition coupled with capillary electrophoresis and laser-induced fluorescence detection. In human carcinoma cells, glycols induced by 0.25 Gy could be detected. In addition, an inducible repair process for radiation-induced damage to DNA bases was reported. Comparisons of effectiveness of bd repair among species have not been conducted, even for the overall level of bd, measurable as endonuclease sensitive sites. Some indirect evidence that bd repair can vary among species has been presented by Heartlein and Preston (1985) who used chromosome aberration frequency as a surrogate for DNA repair. It is quite plausible to expect a difference in repair of bd among species given the several enzymatic steps necessary for the excision process, and the fact that the excision repair of UV-induced DNA damage is known to vary significantly among species (Hart and Setlow, 1975). Again, to understand the mechanisms underlying relative radiosensitivity for a variety of endpoints among species, or among different cell types for a single species, it will be necessary to unravel the process of bd repair for a range of specific types of bd. Understanding the excision-repair process for ionizing radiation-induced DNA damage lags somewhat behind that for UV irradiation.

It should further be emphasized that the majority of information on DNA repair has been obtained from *in vitro* exposure in proliferating cells, often using transformed cells. It is not readily possible to extrapolate to *in vivo* situations for any species, or from transformed to normal cells, and generally not from cycling to senescent cells. It is now being appreciated to a greater extent than ever before that the complexities of the major cellular processes (replication, division and repair) can make for considerable difficulties in extrapolation. In addition, recent research highlights the interrelationships among these cellular processes and the dependence of one upon the others.

5.2.3 Characterization of Genes (Enzymes) Involved in DNA Repair

The intention of this Section is not to provide a detailed description of the isolation and characterization of DNA repair genes, or of the genes and putative enzymes themselves. Rather, the goal is a general description of the types of enzymes so far characterized and how such information might relate to the ability to extrapolate endpoint sensitivity from species to species.

Until quite recently, much of the information on the molecular characterization of DNA repair genes has been for nucleotide excision repair processes, making use of cells from individuals with xeroderma pigmentosum and yeast and mammalian cell mutant strains, for genes involved in specific chemical DNA adduct repair (*e.g.*, methylguanine methyltransferase and DNA glycosylases) and for mismatch repair. As far as mammalian genes are concerned, the majority of the information has been obtained for humans, although clearly with the extent of DNA sequence homology among a range of mammalian species, it will be a relatively simple matter to isolate and characterize repair genes from other species.

The nucleotide excision repair process has been extensively studied and a complete characterization of the genes and proteins involved has been obtained (Aboussekhra *et al.*, 1995; Sancar, 1995; Wood *et al.*, 2001). During the development of the model, it was established by Selby and Sancar (1993) that the nucleotide excision repair process had a link to transcription. It rapidly became clear that many of the genes mutated in individuals with xeroderma pigmentosum, and UV-sensitive yeast strains and mammalian cell lines are, in fact, components of the transcriptional apparatus, specifically the transcription factor TFIIH (Drapkin *et al.*, 1994; Schaeffer *et al.*, 1994). The strand incisions themselves involve the transcription complex, and this probably dictates the specificity of the incision (Chalut *et al.*, 1994; Sancar, 1994). It might be expected that nucleotide excision repair would be quite similar across species since it represents a modification of transcription that is similar in its components and fidelity across species. However, as noted above, there is some evidence that this similarity of repair efficiency is not observed across a range of mammalian species (Hart and Setlow, 1975), but it might be expedient to readdress this issue since more sophisticated molecular methods are now available.

It is only in the past 5 or 6 y that progress has been made in identifying genes involved in the recognition and repair of ionizing radiation-induced DNA damage. With this improved understanding, a better conceptual basis for the mechanisms behind ionizing radiation sensitivity has been obtained (Jackson, 2002).

There has been progress in understanding the role of a DNA-dependent protein kinase in DNA repair *via* NHEJ (Gao *et al.*, 1998; Smider *et al.*, 1994; Taccioli *et al.*, 1993). This enzyme consists of two parts, a catalytic component (DNA-PK$_{cs}$) and a DNA binding component. This latter component has been named Ku protein, and is a heterodimer of Ku70 and Ku80 (more accurately designated as Ku86 based on molecular weight). A significant feature of Ku is that it binds to DNA ends, and recruits the DNA-PK$_{cs}$ protein to form a complex. Substrates for the kinase activity of the complex include *p53*, Sp1 (a transcription factor), and RNA polymerase II (Dvir *et al.*, 1992; Gottlieb and Jackson, 1993; 1994). The precise role of the kinase in DNA repair has not yet been established but several intriguing observations have mechanistic implications. A series of rodent cell lines that are x-ray sensitive and deficient in repair of dsb, have parallel deficiencies in the V(D)J recombination of lymphocyte antigen receptor genes. One of these, *Xrs6*, is complemented by *XRCC5* that encodes the *p86* subunit of Ku (Taccioli *et al.*, 1994). *Xrs6* cells are not only defective in Ku but also in the DNA-dependent protein kinase activity (Finnie *et al.*, 1995). These observations describe a link between Ku DNA-dependent activities with the repair of dsb and V(D)J recombination. This association was further extended by the studies of Blunt et al. (1995), who showed that x-ray sensitive V3 mutant hamster cells and homozygous *scid* mice retained Ku activity but were defective in the DNA-PK$_{cs}$ catalytic component of the DNA-dependent protein kinase. Thus, DNA damage in the form of dsb can be recognized by Ku, which then recruits DNA-PK$_{cs}$ to form a kinase complex that can initiate DNA repair through phosphorylation of additional repair proteins, possibly including components of the V(D)J recombination process (Jackson, 2002).

Details of the homologous recombination DNA repair pathway are also being uncovered. This effort has been facilitated by the finding that several human radiosensitivity disorders are the result of defects in double-strand repair *via* homologous recombination (Thompson and Schild, 2002). For example, the *Nbs1* gene that is mutated in Nijmegen breakage syndrome has been shown to be essential for homologous recombination dsb repair (Tauchi *et al.*, 2002). This gene is part of a complex of proteins, Mre11-Rad50-*Nbs1*; it appears feasible that this complex as a whole is involved in dsb repair. It is of additional interest that *BRCA1* and *BRCA2*, genes, which are mutations and increase susceptibility to breast cancer, are strongly implicated in homologous recombination repair of dsb (Thompson and Schild, 2002).

A further significant advance in our understanding of cellular responses to ionizing radiation occurred when, after extensive efforts on the part of a number of research groups, a gene, *ATM*, that is mutated in the autosomal recessive disorder A-T was identified by positional cloning on chromosome 11q22-23 (Savitsky *et al.*, 1995). Its 12-kb message made it refractory to cloning. Despite the fact that four complementation groups had been identified as possibly representing different genes, the *ATM* gene is mutated in all of these. The predicted *ATM* product is similar to several yeast and human phosphatidylinositol-3-kinases that are involved in mitogenic signal transduction, meiotic recombination and cell-cycle control (Keith and Schreiber, 1995). The potential role of A-T heterozygosity in an increase in tumor incidence remains to be determined. Further studies have begun to characterize the role of *ATM* in DNA repair and radiosensitivity (Khanna *et al.*, 2001; Rotman and Shiloh, 1998; Zhang *et al.*, 1998). For example, it appears that *ATM* regulates multiple cell-cycle checkpoints as well as regulating DNA repair and apoptosis. Thus, it is a central regulator of responses to DNA dsb (reviewed in Khanna *et al.*, 2001). The isolation and characterization of additional DNA repair genes will surely help to clarify the description of how a cell recognizes DNA damage and removes it with fidelity or in an error-prone manner.

5.2.4 *DNA Repair and Cell-Cycle Progression*

As will be apparent in subsequent sections, DNA damages induced in G_0, G_1, or G_2 can be converted into chromosome alterations and point mutations as a result of errors in the repair process. This misrepair can be a consequence of joining incorrect DNA ends together during ligation, recombination or excision repair, or from the insertion of an incorrect base during these various repair processes. Teliologically, there is very little that the cell can do about this, given that repair of ionizing radiation-induced DNA damage is a potentially error-prone process, and that there is a cellular need to repair DNA damage.

There is something, however, that the cell can do about the control of entry into the two critical phases of the cell cycle (S and mitotic/meiotic division) with the genome as intact as possible. This is achieved through the development of the so-called "cell-cycle checkpoints" that arrest cells prior to entry into the S phase or prior to the commitment in G_2 to contract and segregate mitotic chromosomes. The signature gene for describing the G_1 checkpoint is *TP53*, a tumor-suppressor gene, with several linked functions that contribute to this general phenotype. It appears that wild-type *p53* protein can inhibit cell-cycle progression by binding to the TATA binding protein that is a component of the transcription complex, thereby inhibiting transcription of genes that are necessary components of cell-cycle progression (Seto *et al.*, 1992). It can also activate the transcription of genes that have a *p53* responsive element (Kern *et al.*, 1991), and act as a regulator by binding to the replication protein RPA (Dutta *et al.*, 1993). Further, *p53* induces the expression of *p21*^{Cip1}, a cell-cycle inhibitor of cyclin-dependent protein kinases, thereby preventing the induction of DNA replication (Harper *et al.*, 1993; Waga *et al.*, 1994; Xiong *et al.*, 1993). Any one or a combination of these features of *p53* could result in cell-cycle arrest.

Perhaps the most important characteristic of *p53*, however; is that its level of expression is enhanced by DNA damage, largely as a result of increased stability caused by post-translational modifications. This then provides a way in which its checkpoint function can be available as needed. Coupled to the checkpoint function, evidence is accumulating that wild-type *p53* can bind to single-stranded DNA ends, at the sites of ssb or dsb and, in bacteria, can catalyze DNA renaturation and strand transfer (Bakalkin *et al.*, 1994). This evidence suggests that *p53* plays a direct role in the repair of DNA breaks.

Further data support this contention. It has been suggested that not only does *p53* stimulate the synthesis of *p21*, it also upregulates the expression of Gadd45 (Smith *et al.*, 1994). However, the role of this upregulation of Gadd45 in nucleotide excision repair is a matter of debate (Kazantsev and Sancar, 1995). Both Gadd45 and *p21* can complex with proliferating cell nuclear antigen; *p21* seems to prevent the proliferating cell nuclear antigen from conducting replication of long stretches of DNA, but not the short stretches involved in repair (Li *et al.*, 1994). The cell cycle is checked and the DNA damage repaired through the coordinated activity of *p53*. Thus, *p53* can be considered as being involved in preventing the cell from progressing into the DNA replication phase, and at the same time using this checking time for the repair of induced (endogenously or exogenously) DNA damage. The way in which wild-type *p53* can carry out these functions, and why the various mutant *p53*s cannot, is elegantly demonstrated by the crystal structure of the tumor-suppressor DNA complex (Cho *et al.*, 1994).

The majority of recovered mutations, as might be suspected, occur in the core domain, which contains the sequence-specific DNA binding activity of the *p53* protein. A role for *p53* in G_2 prior to mitosis has not been established, although it has been reported that there is either no *p53*-induced cell-cycle checkpoint in response to DNA damage in G_2, or that there can be in certain circumstances (Bunz et al., 1998). However, it is quite feasible that there is a *p53*-mediated DNA repair function that is active in all stages of the cell cycle (Donner and Preston, 1996). It could be that this repair involves *p53*-binding at DNA strand-breaks as a signal for repair enzymes, much as the stalled RNA polymerase transcription-coupled repair factor can serve in this capacity for the nucleotide excision repair system in *E. coli* (Selby and Sancar, 1993) or the TFIIH transcription complex and associated proteins in eukaryotes (Sancar, 1994). There is also what might be described as a "bail out" process, namely apoptosis, or so-called programmed cell death, that involves *p53* protein expression (Clarke et al., 1993; Lowe et al., 1993).

In simple terms, if a cell contains so much DNA damage that check and repair would be ineffective, the cell enters the apoptotic pathway. Cell death in this case would be preferable to high probability of mutation. It should also be noted that in lymphoma cells or activated T cells, apoptosis can be induced following genotoxic exposures of *Trp53 –/–* mice (Strasser et al., 1994). Clearly, the processes of cell-cycle checking and/or apoptosis in response to induced DNA damage are complex and certainly can require more than the activities of *p53*. However, *p53*'s mode of action serves as an example of how cellular gene expression can influence the sensitivity to radiation-induced genetic alterations, and how the relative sensitivities of different cell types within a species or among different species could be influenced by the specific genotype for a whole gamut of regulating genes.

Perhaps the simple view of cell-cycle checkpoints being a singular way by which cells can reduce the frequency of mutation in an exposed cell is indeed too simple. For example, Pardo et al. (1994) showed that transfection of rat embryo cells *in vitro* with mutant *p53* can increase the intrinsic radiation resistance as measured by cell survival. The mechanism whereby this occurs is currently unclear. Similar results have been reported by Lee and Bernstein (1993) with mouse hematopoietic cell lines. It has also been shown in a number of experiments that oncogenic sequences such as *Ras* can increase the radiation resistance of a number of tumor and normal cell lines (McKenna et al., 1990; Samid et al., 1991). The activities of many if not all activated oncogenes are as growth factors that influence the orderly progression of the cell cycle or the entry into or exit from the cell cycle. General reviews can be found in the comprehensive book by Murray and Hunt (1993) and in Munger (2002), although these are not completely current because of the nature of progress in this research area. Similarly, a brief resume of the role of cyclins and cyclin dependent protein kinases in the initiation of DNA replication and mitosis can be found in Lock and Wickramasinghe (1994). The overview presented here is intended to provide an indication of the complexity, not a detailed analysis, of the control of the cell cycle. It serves to highlight how differences in the control processes among species, particularly in response to DNA damage, could greatly influence sensitivity to mutation induction. As noted by Gard and Kropf (1993), "the goal of the cell cycle is not to regulate itself, but rather to regulate all cellular processes required to ensure that two daughter cells are faithfully reproduced by cell replication and division." Changes in these cellular processes will ultimately be involved in the development of tumors. Identifying the specific changes that can be induced by ionizing radiations, and how these can be involved in tumor formation will provide a basis for making extrapolations of risks for specific tumor types in animal models to humans, even if merely to provide a sounder basis for the pragmatic approach.

5.2.5 *Genetic Susceptibility to Ionizing Radiations*

The preceding section on ionizing radiation-induced DNA damage repair, and the variation of DNA repair kinetics in the different stages of the cell cycle, makes it fairly clear that mutations in the genes involved in these cellular processes can strongly influence susceptibility to radiation-induced genetic alterations.

For DNA repair, alterations in DNA-dependent kinase can greatly influence DNA repair fidelity through recognition of DNA damage and the subsequent repair itself. The fact that rodent cell lines and *scid* mice share defects in Ku and/or DNA-PK$_{cs}$ suggests that rodents and humans will be similarly susceptible to the consequences of ionizing radiation-induced DNA damage. In this context it will be interesting to determine if a similar process of recognition and repair is operational for DNA damage induced by high-LET radiations, given some differences in structure of the induced DNA dsb.

Individuals who are homozygous recessive for the A-T gene (*ATM*) are x-ray sensitive for cell killing and have an increased susceptibility to develop leukemias and lymphomas (IARC, 2001). The frequency of the A-T gene is estimated to be 1 in 40,000 to 1 in 200,000. However, the frequency of the heterozygote is ~1 %, and it has been argued that an increased susceptibility and/or sensitivity of such individuals to cancer induction could account for a significant fraction of human cancers, up to 8 % of breast cancers. The epidemiological data that have been used to support this contention, for obligate heterozygotes (Swift *et al.*, 1991), has been the subject of considerable controversy (reviewed in Easton, 1994; IARC, 2000; ICRP, 1999). A particularly pertinent discussion involving a number of prominent research scientists can be found in the New England Journal of Medicine (NEJM, 1992). More recent epidemiological studies do not support this interpretation (Angele and Hall, 2000; Appleby *et al.*, 1997; Chen *et al.*, 1998; Lavin, 1998). The data from Scott *et al.* (1994; 1996) are potentially interesting in this debate. These authors showed that sensitivity to x-ray-induced chromatid breaks in G_2 lymphocytes was greatest in A-T –/–, least in A-T +/+ with A-T +/– being intermediate. Of particular interest was the fact that among breast cancer patients, 42 % showed an aberration frequency that was similar to the A-T heterozygotes. This does not mean that this proportion of breast cancer patients are A-T heterozygotes but rather that predisposing genes involved in processing DNA damage are more likely to be present in breast cancer patients and perhaps causative in part. Hall *et al.* (1998) showed that ataxia heterozygotes are disproportionately represented among prostate cancer patients who experience severe stochastic effects after radiotherapy.

It is important to note at this point that genetic predisposition to cancer and sensitivity to radiation-induced cancer have been largely restricted to discrete genetic subgroups because of the ease of their detection (Chakraborty and Sankaranarayanan, 1995). Any effects of radiation even in those groups are likely to be detectable only for high (therapeutic) doses (ICRP, 1999). In the context of population effects, it is perhaps the more subtle, genetically controlled increases in radiation sensitivity (*i.e.*, genetic poly morphisms) that will be of greater impact (ICRP, 1999). These types of subtle changes should be the subject of enhanced study.

It has been proposed by Sanford and colleagues (Parshad *et al.*, 1983; Sanford *et al.*, 1989) that increased sensitivity to the induction of chromatid breaks and nonstaining gaps following x-ray exposures in G_2 is an identifier of cancer-prone individuals. The aberration sensitivity would be a measure of DNA repair deficiency. A particular concern is that for all the cancers, and other syndromes assessed, there was a similar enhancement in sensitivity. This is unlikely, and in this regard Scott *et al.* (1996) have been unable to repeat the Sanford *et al.* assay, except for the A-T homo- and heterozygotes. This type of assessment clearly requires further study. It is expected that alterations in DNA repair would be involved in sensitivity to tumor formation for some tumors, and that in some cases this would involve repair deficiencies for ionizing radiation.

There is little or no information on a genetic sensitivity to tumor formation by high-LET radiation, largely because of a lack of study. Studies with A-T homozygotes show that the cellular radiosensitivity observed with x rays is not detectable with alpha particles (Coquerelle *et al.*, 1987; Lucke-Huhle *et al.*, 1982). This probably reflects differences in the types of DNA damage induced and the mode of repair for low- and high-LET radiations (Stenerlow *et al.*, 2002).

The subject of the impact of predisposition to radiation-induced cancer has been addressed by an ICRP task group (ICRP, 1999). This report shows that the impact is quite low at the population level, although it can be very significant at the individual level. The role of genetic susceptibility in cancer induction will remain a rapidly growing field as our understanding of the molecular basis of cancer continues to expand. The proceedings of a Cold Spring Harbor symposium serves to exemplify this (CSH, 1994), and articles by Hahn and Weinberg (2002) and Hanahan and Weinberg (2000) support this viewpoint.

5.2.6 *Conclusions*

Section 5.2 of this Report summarizes the current state of our knowledge on the induction of DNA damage by radiations of different qualities. It appears that dsb and some bd, quite possibly at multiply damaged sites (or sites of clustered damage), are the most important for the induction of chromosomal alterations and point mutations. These genetic endpoints are largely the consequence of misrepair during one of the several known DNA repair processes, although errors of DNA replication can occur for DNA damage remaining at the time of replication. A number of cellular components and functions are involved in ensuring efficient and accurate repair. Mutations in one or more of these processes will result in sensitivity to the induction of genetic damage. The induction of DNA damage is expected to be similar across species, with the frequency being proportional to DNA content. The repair of DNA damage can vary across species and for specific genotypes, but the extent of this variation is somewhat uncertain, although probably not large enough to preclude pragmatic extrapolation across species.

5.3 Effects of Ionizing Radiations at the Cellular Level

5.3.1 *Point (or Gene) Mutations*

Gene alterations involving single base pair changes, either transitions, transversions, frame shifts, or deletions, can arise at the sites of dsb or bd. They will be generated during ligation or the resynthesis step of excision or recombination repair, or from errors of replication of a damaged template during the S phase of the cell cycle. On the basis of this mechanism of formation, the dose-response curve is predicted to be linear for acute and chronic exposures to low- and high-LET radiations (Preston, 1992). In addition, the relative sensitivity will be determined by the size of the target, the efficiency and fidelity of repair, the nature of the DNA lesions (*i.e.*, as influenced by the quality of the radiation) and the dose rate. The latter is of importance when errors of replication are involved, since the relationship between induction of damage, repair, and time of replication will be different for different rates of damage formation.

How will these several sources of variation influence relative mutation frequency among different species? The size of the target is unlikely to vary much for oncogenes and tumor-suppressor genes, for example, from one species to another, and so this is unlikely to be a variable to be considered when extrapolating for a specific gene. However, the overall mutation frequency will depend upon DNA content of the cell, and this can vary among mammalian species by ~10 % for diploid cells (Bachmann, 1972). This does not appear to represent a significant modifier of relative sensitivity, especially when it is considered that the mutations of relevance will be in coding or transcription control regions (*e.g.*, promoters and enhancers), whereas the largest variations in genome size are for repetitive DNA sequences. It seems reasonable to predict that among species the rate of repair of induced DNA damage would be the variable and thereby, influences sensitivity to radiation-induced mutations the most. This would result in different amounts of DNA damage being present at the time of replication (*i.e.*, the slower the repair, the higher the mutation frequency). The fidelity of repair also would influence sensitivity to mutation induction, and this fidelity is predicted to be lower for the more extensive damage produced by high-LET radiations.

It is difficult to obtain information from the published literature that allows for an interspecies comparison of radiation-induced point mutations. Data collated by Sankaranarayanan (1991b) indicate that for a single locus, the *HPRT* gene, the mutation frequency per gray for human, Chinese hamster CHO or V79, and mouse lymphoma cells is ~1 × 10^{-5}. This figure includes deletions as well as point mutations, but it allows for a reasonable conclusion that there is likely to be little difference for point mutations alone. How this links to DNA repair kinetics and fidelity is difficult to determine since data for repair of ionizing radiation-induced DNA damage across species is lacking. In a somewhat circular way, since the induced mutation frequencies (as well as the spontaneous frequencies) are quite similar across species, it is quite likely that the repair kinetics and fidelity are also similar. However, this relatively broad generalization needs to be tempered by the fact that there are exceptions by which mouse and human data do not diverge (UNSCEAR, 2001).

Some caution is needed in accepting the UNSCEAR (2001) generalization. For instance, Grosovsky and Little (1985) showed that the mutagenic risk in TK6 human lymphoblast cells *in vitro* to low, daily-repetitive doses of x rays can be accurately estimated by linear extrapolation from high-dose effects. Rodent cells *in vitro*, however, generally display a curvilinear response but show responses similar to those of human cells *in vitro* when exposed to an acute single-dose exposure (*e.g.*, Burki, 1980; Fox, 1975; Jostes *et al.*, 1980; Nakamura and Okada, 1981; Thacker and Stretch, 1983). The overall results are complicated because, for instance, stationary G_1 cells have a different response than stationary cells allowed a 5 h holding period (Thacker and Stretch, 1983), and different stages of the cell cycle have different sensitivities to lethality and mutation induction (Burki, 1980).

From the foregoing discussion, it would appear to be reasonable to conclude that the radiation-induced point mutation frequency for a specific gene will be similar across species, on the assumption that the gene size does not vary significantly among mammalian species. This allows for an expanded conclusion that a specific step in the multistage process of tumor development, when it involves a point mutation in the same gene, will have a similar mutation frequency in different mammalian species. Extrapolation will be relatively straightforward in such cases. It still remains to be determined how often this particular mode of extrapolation can be made.

5.3.2 *Chromosome Aberrations and Deletion Mutations*

All types of chromosome aberrations induced by ionizing radiations also arise as background events. It is generally agreed that all types arise in G_1 or G_2 phases of the cell cycle by errors of repair, either failure to complete repair for terminal deletions and incorrect repair for exchanges and interstitial deletions. For cells in the S phase of the cell cycle, replication of unrepaired DNA damage can lead to errors in the form of chromosome aberrations, and aberrations can also arise from repair errors in the S phase prior to replication. In terms of the relative sensitivity across species of genetic events that are related to cancer induction, it is appropriate to restrict the discussion to the more pertinent transmissible ones, namely reciprocal translocations, fairly small interstitial deletions, and some inversions. The great majority of dicentrics, rings and terminal deletions are cell lethal as a result of interference with chromosome segregation at anaphase or loss of large quantities of genetic information in the form of acentric fragments. It has long been suggested and more recently demonstrated that the frequency of induced dicentrics is equal to the frequency of induced translocations (Lucas *et al.*, 1992). This allows for conclusions based upon the analysis of dicentrics to be interpreted in terms of translocations. However, the development of fluorescence *in situ* hybridization techniques for painting whole chromosomes has allowed direct measurement of translocations in somatic cells to be made much more readily (Lucas *et al.*, 1995).

Since the formation of chromosome alterations involves the conversion of induced DNA damage into an aberration through errors in DNA repair or replication, then any differences in kinetics or fidelity in these processes among species could lead to differences in sensitivity to ionizing radiations.

Although it is required that both halves of a DNA double helix be involved in the formation of a chromosome aberration, it is not necessary for the dsb to be induced by direct ionizations at the DNA level. It is possible to produce double-strand interactions through bd repair *via* OH-radical interactions with DNA, and possibly through misrepair of ssb. The latter is unlikely, given the high probability of correct rejoining of a ssb, from considerations of chromosomal (DNA) geometry and the very rapid repair of ssb, thereby limiting the time available for interactions.

It is straightforward to perceive how directly-induced dsb can lead to chromosome alterations; unrepaired dsb would result in deletions and misrepaired dsb in exchanges (inter- and intrachromosomal). This has been formulated in the breakage first model for aberration formation, originally proposed by Sax (1938). This model further predicts that the dose-response curve for low-LET-induced inter- and intrachanges requiring two breaks (one or two tracks) for their nonlinear formation, follows the general formula

$Y = aD + bD^2$, whereas that for terminal deletions, requiring only one break (one track) is linear. For high-LET radiations, the breaks (one or two) required for aberration formation are presumed to be produced by one track, and so dose-response curves for all aberrations will be linear. For chronic exposures to low-LET radiations, the frequency of induced dsb per unit time will be considerably lower than that for an acute exposure, and, with rapid repair (<5 h), the probability of producing two-track aberrations will be very low. The outcome is that the dose-response curve for inter- and intrachanges will be linear ($Y = aD$), with that for terminal deletions being unchanged. It is difficult to confirm the latter prediction since the frequency of terminal deletions in mammalian cells is low. For high-LET radiations and for all aberration types, there is no effect of dose rate on the frequency of chromosome aberrations formation.

Despite the fact that the breakage first hypothesis and the sole involvement of dsb in chromosome aberration formation has simplicity, it fails to explain the available data, and does not allow for the involvement of a high proportion of the induced DNA damage, most notably alterations to bases (bd). In addition, more recent data have shown that chromosomal alterations and gene mutations can be induced in cells that are not directly traversed by a radiation track (the so-called "bystander effects"). This observation is discussed in more detail in Section 5.3.5.1.

The model described by Revell (1974) proposes that all chromosome aberrations, including deletions, are produced from an exchange process, essentially complete for inter- and intrachanges and incomplete for deletions. In its original concept the exchanges were derived from interactions of lesions. It has become clear that the lesions envisioned by Revell can be DNA dsb, base alterations, or multiply-damaged sites, and that the interaction to form an aberration can occur at the time of DNA repair (G_1, G_2 and much of S phases of the cell cycle) and of DNA replication (S phase). The relative involvement of repair and replication in the S phase in the formation of aberrations will be dependent on the time between damage induction and replication of a particular genomic region, *i.e.*, damage induced some hours before replication will be repaired prior to replication, and, as a corollary, damage induced in S postreplication will not be subject to replication errors.

On the basis of the published literature, it seems that a combination of the two models (breakage-first and exchange) is operational. The study by Duncan and Evans (1983) perhaps best represents this view. They showed that a proportion (~20 %) of chromatid deletions induced by bleomycin (which is radiomimetic) were associated with sister chromatid exchanges. Revell's exchange hypothesis predicts that 40 % would be, whereas the breakage first model predicts that 0 % would be. There is more than a "wish-to-know" reason for being able to describe the specific mode of formation of chromosome aberrations, because the factors that could influence sensitivity among cell types and/or species appear to be dependent upon the mechanism, particularly the DNA repair processes, and whether one or two tracks are likely to be involved in formation at low-LET (one or two lesions at high-LET).

An additional example of the relevance of mechanism to interspecies extrapolation issues is provided by the studies of Brewen and Brock (1968). Using cells from a wallaby, that has a simple karyotype, they showed that the great majority of chromosome-type terminal deletions, induced in G_1 by x rays, were the result of incomplete exchange formation. This means that their formation would be dependent upon the repair characteristics of pairs of DNA lesions able to interact, *i.e.*, a coincidence in time and a proximity in space. It is frequently assumed that terminal deletions are the quintessential example of a single DNA break, an assumption that appears to be incorrect.

In summary, all classes of chromosome aberrations arise from interactions between pairs of DNA damages (strand breaks and damaged bases) and that interchanges (dicentrics and reciprocal translocations), interstitial deletions, and rings represent misrepair events, and that terminal deletions represent incomplete repair. Thus, extrapolation will require a knowledge of the repair kinetics in different species for different classes of DNA damage.

Griffin *et al.* (1996) assessed the efficiency of 1.5 keV aluminum x rays at inducing complex chromosome aberrations. The authors offered as one interpretation that damaged DNA might interact with undamaged DNA to produce the aberrations. This is in contrast to the data of Cornforth (1990) who concluded that a one-hit exchange did not occur. The importance of a one-hit exchange process to dose response is readily apparent, and the question of its likelihood, especially at low doses, requires further study.

Relatively few appropriate data are available to derive relationships between rate of DNA repair and chromosome aberration frequency. Data of Heartlein and Preston (1985) show that there is a relationship between rate of repair of DNA bd and frequency of x-ray-induced chromosome aberrations for human, marmoset, pig and rabbit lymphocytes. The relationship is that the faster the repair, the higher the aberration frequency, leading to the conclusion that rapidity of repair leads to a higher probability of coincident bd that can interact to produce aberrations. The differences in frequency for neutron-induced aberrations were significantly less and could be explained by differences in DNA content. It can be argued that extrapolation across species for high-LET radiations is mostly dependent upon relative DNA content, probably due to the fact that DNA dsb (or multiply damaged sites) are the lesions involved in aberration formation, and the two lesions needed for aberration formation are produced by one track.

Preston (1992) has argued that the dose response for chromosome aberrations induced by low-LET radiations can be viewed not only as a combination of one- and two-track events ($Y = aD + bD^2$), but also that the two components are representative of different DNA damages; the linear being from dsb and the dose-squared from bd. The implication is that at low doses the major event in the formation of aberrations will be misrepair of dsb. Similarly, for low-dose rates, the linear curve obtained for aberrations is likely to be representative of dsb interactions. Thus, extrapolating across species for low doses and dose rates will require consideration of kinetics and fidelity of dsb repair. There is little evidence to suggest that there are major differences among different species, at least for cells *in vitro*. However, it is necessary to note that a definitive set of studies is not available. It has been reported by Frankenberg-Schwager and Frankenberg (1990), for example, that repair of x-ray-induced dsb is very similar in yeast and mammalian cells when DNA content is taken into account. The modes of repair are similar and are discussed in Sections 5.2.2.1 and 5.2.2.2 above.

For high-LET radiation-induced aberrations, the dose-response curve is linear over the dose range studied, and presumably also at low doses, leading to the conclusion that DNA double-strand lesions are involved in their formation and that interacting pairs of these lesions are produced by one radiation track (Preston, 1992). The molecular nature of the lesions has neither been established nor has the mode of their repair, although some form of recombination repair is most likely. On the basis of the argument developed above with low-dose rates of high-LET radiation, there would be little or no reduction in aberration frequency compared to those at high-dose rates (Preston, 1992). Thus, extrapolation across species for chromosome aberration induction by high-LET radiations, irrespective of dose or dose rate, can be estimated from ratios of DNA content. In addition, it seems possible that at least for high-LET exposures, there might be an influence of bystander effects (Section 5.3.5.1) that could be differentially expressed in different species.

5.3.3 *Use of Mechanistic Data on Mutation and Chromosome Aberration Induction*

The introduction of molecular biology techniques into the field of radiation biology, together with functional assays of response, has led to a clearer understanding of the mechanism of induction of mutations and chromosome aberrations by different ionizing radiations. This by itself is of value, but it takes on a greater significance when considering the methods for extrapolating cancer risks across species. The reason for this is that tumor development is clearly in large part under genetic control, with some steps in the multistep process being associated with specific genetic alterations. The best example of the multistep process is colon cancer (Fearon and Vogelstein, 1990) where specific changes for each of the five stages have been identified. The involvement of a mutation in DNA mismatch repair in hereditary nonpolyposis colorectal cancer (Fishel, 2001; Fishel *et al.*, 1993; Leach *et al.*, 1993) also serves as an excellent example of the involvement of mutator phenotypes or cell selection that can enhance the probability of subsequent mutations in the progression pathway being produced. The steps can involve specific gene mutations, deletions, translocations and aneuploidy (Sandberg, 1993). Thus, knowing or predicting the relative sensitivities for the induction of these types of genetic alterations across species compared with that for humans will provide a factor for predicting relative sensitivity for tumor formation. This will provide a basis for confidence in estimating human cancer risk from data obtained in laboratory animals, and would, in turn, give greater or lesser confidence to a simple pragmatic approach.

A somewhat different description of the cancer process that moves away from the initiation, promotion, progression paradigm was described by Hanahan and Weinberg (2000). Their approach allows for a very different process of extrapolation across species that is phenotype-based rather than genotype based. An extensive review of the tumor database led Hanahan and Weinberg (2000) to propose that tumors, irrespective of species or tissue site, required the acquisition of six characteristics related to the ability of transformed cells to overcome signals that normally limit growth (*e.g.*, cell-cycle control, apoptosis, and angiogenesis). As an extension of this hypothesis, Hahn and Weinberg (2002) described the so-called rules for making human tumor cells. The development of a similar set of rules for tumor formation in other species, the mouse, for example, would allow for specific endpoints for extrapolation to be identified. In light of the very rapid development of genomic and proteomic methodologies and databases, cellular markers to aid in this extrapolation should soon be identified or in fact have been identified in a few cases (Jackson-Grusby, 2002).

5.3.4 *Cell Killing*

For considerations of the relevant cellular data for extrapolation of cancer incidence across species, it is appropriate to include interspecies differences in radiation-induced cell killing. Cell killing is influential on the recovery of mutant cells in the initiation/progression stages of tumor development, and significant differences in cell killing will impact upon relative tumor frequency.

It is interesting to note that, in general, all mammalian cells *in vitro*, normal or malignant, have a similar sensitivity to cell killing as assessed by D_0 (1 to 2 Gy for x rays). The D_0 for high-LET radiations is lower than that for low-LET (Hall, 2000). The exceptions to this general observation are cell lines derived from known radiosensitive individuals, such as those with A-T. There is considerably less of a sensitivity difference between A-T and normal cells with high-LET radiations (Cox, 1982). This probably reflects a difference in mechanism of induction of cell killing by the different radiation qualities (Preston, 1992). While it has often been argued that *in vitro* responses are not necessarily representative of those *in vivo*, for cell survival (*i.e.*, reproductive integrity) the *in vivo* and *in vitro* parameters are, in general, quite similar.

The D_0 essentially describes the dose required to produce one inactivating event per cell. This component is quite similar for a broad range of cell types. However, the extrapolation number (n) that describes the magnitude of the shoulder on a survival curve can vary significantly among cell types. The survival curves for a number of mammalian cell types have a large shoulder with an n of the order of 5 to 10 for irradiation with x rays. In contrast, primary human cells have little or no shoulder (Hall, 2000). This could reflect differences in DNA repair competence or be reflective of differences in other cellular processes that are involved in cell killing. It is perhaps interesting to note that genetic alterations involved in cell killing (asymmetrical chromosome aberrations and dominant lethal mutations) are induced either by one ionization track or by two tracks, leading to an induction curve that would be the complement of a shoulder-type survival curve. Thus the curve for human cells is an anomaly in this regard, based on the reasonable assumption that the same mechanism of cell killing holds for the different species. The survival curves for high-LET radiations do not have shoulders, indicative of the one-track induction of genetic alterations involved in cell killing.

In general terms, the extent of cell killing among a range of mammalian species, as described by D_0, differs by about a factor of three for low-LET radiations, and given the genetic basis for cell killing, it is reasonable to conclude that this range of about three could hold for genetic alterations for most cell types. There certainly will be some exceptions (*e.g.*, the developing oocyte). This conclusion provides a reasonable level of support for using a pragmatic approach for tumor extrapolation across species.

5.3.5 *Potential Confounders of Dose-Response Curves*

The major aim of extrapolating from data in laboratory animals to humans is to predict responses at low doses for tumors for genetic indicators of tumors. The current models for extrapolation rely almost exclusively upon the assumption of low-dose linearity across the range of responses (NCRP, 2001b). Several recent observations have led to a reconsideration of low-dose linearity and these will be briefly discussed in the following three sections.

5.3.5.1 *Bystander Effects.* The bystander effect is described as a response in cells that are not directly traversed by a radiation ionization track. This response can be genetic or epigenetic. The majority of these bystander responses have been described for high-LET exposures since it is readily feasible to target a small fraction of the cell population (Little, 2003). For low-LET radiations, targeting single cells is much more difficult, and here the majority of bystander effects have been described using tissue culture medium transfer from irradiated cells to unirradiated cells (Mothersill and Seymour, 2001). It is proposed that the bystander effect is mediated by cell signaling pathways being induced thereby leading to reactive oxygen species production in bystander cells (Lehnert and Iyer, 2002). These signaling events can be mediated by gap junction intercellular communication, or occur in the absence of direct cell contact (Azzam *et al.*, 2004). The major question is, "what relevance do the *in vitro* cell culture observations have for *in vivo* exposures, where bystander effects are somewhat difficult to predict, especially for low-LET exposures?" The implications for predictions of tumor outcomes at low doses are related to the estimation of the size of the target cell population which includes the irradiated and unirradiated cells that harbor residual damage. However, tumor data from animal studies, since they are disease-based, already account for any bystander effects. Extrapolations from animal to human cellular responses should be cognizant of potential bystander effects and their relative magnitude in different species and among individuals within the same species. Little information on this aspect is currently available.

5.3.5.2 *Genomic Instability.* The development of widespread genomic instability is a hallmark of tumor development. Stoler *et al.* (1999) provide evidence that an early step in sporadic colorectal tumor progression is characterized by several thousand genetic alterations per cell. Cahill *et al.* (1999) propose that a form of Darwinian selection occurs for specific phenotypes in tumor progression. This type of instability is both a cause and a consequence of the cancer process. The type of genomic instability described following radiation exposures is different and much more limited in extent (Little, 1998). No role for this radiation-induced genomic instability in the cancer process has been described.

On the same grounds as mentioned above, extrapolations based on tumor data already incorporate any role of genomic instability. For extrapolations at the cellular level, it remains necessary to understand better any relationship between cancer-related genomic instability and radiation-induced delayed effects. In the context of the present Report, the need is to describe relative responses in animal models and humans.

5.3.5.3 *Adaptive Responses.* An adaptive response to radiation exposures has been described for chromosomal alterations and mutations in cellular systems and *in vivo* (UNSCEAR, 1994; Upton, 2001). The phenomenon is one whereby the frequency of chromosome aberrations is lower by a factor of about two for a small priming dose (*e.g.*, 0.01 Gy) followed by a challenge dose of ~1 Gy compared with that for the challenge dose alone. A number of possible explanations has been proffered, but none has convincingly explained the phenomenon. It is highly variable in cellular (or tissue) systems and in humans, from individual to individual, some showing an adaptive response and others not. Thus, it is not feasible to draw a single conclusion to describe an adaptive response. In addition, there is no definitive evidence for an adaptive response for tumor outcomes. Thus, extrapolations of tumor data from animal models to humans cannot at this time incorporate a component for adaptive response. Certainly, comparisons of underlying mechanisms of an adaptive response in animal and human cellular systems are needed for models based on mechanisms, especially for extrapolation purposes. However, there is no evidence that the relevant adaptive responses are different or are likely to be, among different mammalian species.

5.3 EFFECTS OF IONIZING RADIATIONS AT THE CELLULAR LEVEL / 111

5.3.6 *Genetic Alterations in Tumors in Humans and Rodents*

The level of confidence for extrapolating from tumor frequencies in rodents to the expected frequencies in humans will be enhanced by two factors. First, it is more informative when specific human tumors and the rodent ones being used for extrapolation purposes have the same histopathologies. This is discussed for a selection of tumors in this Section. Second, an additional strengthening of the extrapolation can be achieved if tumors at the same site have similar molecular alterations associated with them. The most complete comparison can be made for spontaneous (background) tumors in humans and rodents with only limited data on radiation-induced tumors in humans and rodents. This Section will provide such comparisons where they exist, but it should be noted that the rodent data are quite limited. The most extensive database is for human spontaneous tumors, with chemically-induced tumors being the most prominent for rodents.

A comprehensive review of the data underlying the approaches for extrapolating across species has been published by Balmain (2001).

5.3.6.1 *Oncogene Activation.* The largest data base for comparisons to be made between human and rodent tumors is for the *Ras* oncogene (Barrett and Wiseman, 1992). The incidence of *Ras* gene alterations is very dependent upon the tumor type in humans (Bos, 1989). Table 5.1 illustrates this point. The frequency of *Ras* mutations can vary from very frequent for pancreatic carcinomas, and lung and colon adenocarcinomas to relatively rare for breast, cervix, melanoma, bladder, liver and kidney, to very seldom or never observed for ovary, stomach and neuroblastoma. It is also of relevance that *Ras* mutations are not essential for tumor development and are generally not a predominant event. There is an extensive data base on the presence of *Ras* mutations in rodent tumors, and a comparison can be made between their incidence in human and rodent tumors (Table 5.2). The incidence of *Ras* mutations is quite similar for some tumors in one or more rodent species and humans, but rarely is there an across the board concordance. For example, K*Ras* activation is very high in human and hamster pancreatic carcinomas but very low in the same tumor in rats (Van Kranen *et al.*, 1991). The frequency of K*Ras* activations is high in mouse and rat lung adenocarcinomas (Stegelmeier *et al.*, 1991; You *et al.*, 1989), with a similar incidence in sensitive and resistant mouse strains (Devereux *et al.*, 1991). For comparison, the frequency in spontaneous lung tumors of mice is 40 %, and in human lung

TABLE 5.1—*RAS mutations in human tumors (Bos, 1989).*

Tumor Type	RAS Positive	Percent RAS Positive
Breast	2/86	2
Cervix	7/106	7
Lung, adenocarcinoma	19/63	30
Colon, adenocarcinoma	119/277	43
Pancreatic	128/156	82
Thyroid	8/15	53
Melanoma	8/50	16
Bladder	7/67	10
Liver	3/31	10
Kidney	3/30	10
Ovary	0/37	0
Stomach	0/33	0
Neuroblastoma	0/25	0

adenocarcinomas it is ~30 % (Mitsudomi *et al.*, 1991; Slebos *et al.*, 1991). This incidence can be altered under various exposure scenarios, being up to 100 % for mice depending upon the inducing agent (Slebos *et al.*, 1990; You *et al.*, 1991).

The incidence of *Ras* gene activation in some rodent liver and mammary cancers can be as high as 100 %, whereas such mutations are very rare in the same types of human tumors (Bos, 1989). However, different mouse strains have very different frequencies of *Ras* mutations in liver tumors. For example, the incidence of H*Ras* mutations is high in the strains that are susceptible to spontaneous and carcinogen-induced liver tumors, and lower in some strains resistant to liver tumor formation (Buchmann *et al.*, 1991). This is not a general rule since a high incidence of H*Ras* mutations has been observed in vinyl carbamate-induced tumors of resistant strains (Stanley *et al.*, 1992). Thus, for liver tumors the pragmatic

TABLE 5.2—*A comparison of Ras and RAS mutations in rodent and human tumors, respectively (Barrett and Wiseman, 1992).*

Tumor Type	Percent Tumors with Mutated *Ras* or *RAS*	
	Rodent (*Ras*)	Human (*RAS*)
Lung	40 – 100 % (A/J and C3H/HeJ mice)	22 – 33 %
Liver	30 – 100 % (B6C3F$_1$ and C3H/HeJ mice) 0 % (C57BL/6J mice)	0 – 10 %
Mammary	20 – 100 % (rat)	0 – 8 %
Pancreatic	95 % (hamsters)	75 – 93 %
Squamous-cell carcinoma (skin)	50 – 90 %	46 %

extrapolation of tumor incidence across species is not enhanced by the molecular data.

A similar scenario holds for rat mammary carcinomas. The incidence of *Ras* mutations is around 80 to 100 % following an acute methylnitrosourea treatment (Zarbl et al., 1985). It has been proposed that *Ras* mutation is an initiating event for rat mammary carcinomas (Kumar et al., 1990). This is very unlikely to be the case for the great majority of human breast tumors, since a very low proportion contain mutated *RAS* (Bos, 1989).

There is minimal concordance for *Ras* involvement with spontaneous and chemically-induced rodent liver and mammary tumors and spontaneous human tumors. However, it is quite clear that there are several different routes for reaching the same endpoint of uncontrolled growth, and these need to be explored further for the various tumor types to establish additional similarities and differences.

There is relatively little information available on the possible role of *Ras* mutations in radiation-induced tumors, but one or two examples serve to illustrate similarities and differences among species and/or strains.

Data for thyroid tumors are presented in Table 5.3. Methylnitrosourea- and x-ray-induced thyroid tumors in rats, with H*Ras*

TABLE 5.3—*Comparison of Ras or RAS acitivation in thyroid carcinomas induced by different carcinogens (rat versus human, respectively) (Barrett and Wiseman, 1992).*

Species	Carcinogen	Gene Activated
Rat	Radiation	KRas
Rat	Methylnitrosourea	HRas
Human	X rays	KRAS
Human	"Spontaneous"	HRAS/NRAS

being predominantly mutated by methylnitrosourea and KRas by radiation (Lemoine et al., 1988). RAS is also mutated in human thyroid tumors, with KRas activation being predominately involved following irradiation, and HRas activation and to a lesser extent NRas activation for "spontaneous" tumors. Thus, for radiation-induced tumors there is a similar oncogene mutation (KRas) in rats and humans. However, between 10 and 50 % of tumors in both species do not contain mutations in any of the Ras genes, demonstrating that other genetic alterations can lead to thyroid tumor formation. Additional information is available for human thyroid tumors that requires comparative studies in rodent tumors if the data are to be used for interspecies extrapolation. About eight oncogenes have been associated with background thyroid tumors. In radiation-induced papillary carcinomas arising following exposure of children as a result of the Chernobyl accident, there were no mutations in RAS, in TSH receptor genes, or in TP53. Of significance was the observation that ~60 % of these tumors contained a rearrangement of RET (Rabes and Klugbauer, 1998). The most frequent of these rearrangements was for RET/PTC3. It has been proposed that a balanced intrachromosomal inversion leads to a fusion of the ELE and RET genes. Results from experiments with RET/PTC1 transgenic mice support a suggested role for this rearrangement in mouse thyroid tumor development (Jhiang et al., 1998). Experiments with RET/PTC3 transgenic mice were reported to be in progress. This information, in general, does provide some increased level of confidence in using direct comparison for rat to human tumors.

The original observation that activated Ras was present in radiation-induced tumors was by Guerrero et al. (1984). Using an *in vitro* transformation system, they showed that the KRas gene was specifically activated in radiation-induced thymomas of (AKR × RF) F_1 mice, but N*Ras* was activated in chemically-induced tumors. Corominas *et al.* (1991) extended these studies to determine the presence of *Ras* activation in gamma-ray- or neutron-induced thymic lymphomas in female RF mice. In summary, they found that the proportion of tumors containing an activated *Ras* gene was quite similar for gamma rays (9/37 or 24 %) and neutrons (4/24 or 17 %). The majority involved K*Ras* mutation with one tumor for each radiation quality containing an N*Ras* mutation. However, it was notable that the mutation spectra were different for the two radiations: seven of nine gamma-ray tumors with a *Ras* mutation had a GGT to GAT change at codon 12, whereas none of the neutron-induced tumors did, although 2/4 did contain mutations at codon 12. This difference in spectrum could reflect differences in the DNA damage induced by neutrons and gamma rays and differences in repair fidelity at different sites along the gene for those damages. In contrast to those data, Belinsky *et al.* (1996; 1997) showed K*Ras* and *p53* pathways were not involved in lung tumors induced in rats by a wide range of agents including x rays.

Another set of data provides some valuable comparisons for the involvement of *Myc* in B-cell tumors (see review by Wiener and Potter, 1993). B-cell tumors are broadly characterized as involving translocations, and in several different kinds of these tumors there is a consistent type of alteration known collectively as *Myc*-associated chromosomal translocations and rearrangements. These *Myc*-associated chromosomal translocations and rearrangements are nonhomologous or illegitimate recombinations involving the domain of the *Myc* gene or its flanking regions and an immunoglobulin gene. Not only do they occur in different B-cell tumors but also in different species, for example, endemic and sporadic Burkitt lymphomas in humans (Pelicci *et al.*, 1986), spontaneous immunocytomas in rats (Wiener *et al.*, 1982), and induced plasmacytomas in mice (Ohno *et al.*, 1984). The translocations directly or indirectly dysregulate the transcription of the *Myc* protooncogene. Thus, although the sites of specific translocations vary among rats, mice and humans, the end result appears to be the same.

There are additional examples of a possible role for changes in *Myc* in tumor formation, including those that are radiation-induced. Niwa *et al.* (1989) reported an overexpression of *Myc* in ~50 % of radiation-induced sarcomas and a range of other radiation-induced tumors in BCF_1 mice. Some radiation-induced mouse osteosarcomas (Schon *et al.*, 1986; Van der Rauwelaert *et al.*, 1988) and thymomas (Bandyopadhyay *et al.*, 1989) also showed overexpression of *Myc*. There was no evidence of *Myc* gene amplification in any radiation-induced tumors, although there was with other inducing agents.

Thus, there is some evidence for induction of overexpression of *Myc* in radiation-induced tumors, and those arising spontaneously or induced by chemical agents in mice, rats and humans, but it is quite difficult to establish a pattern. In part, the data are limited and also they probably reflect different involvement of *Myc* in tumors in different species. Comparisons are made difficult by the lack of similarity of radiation-induced tumors in different species.

Leukemias and lymphomas in humans, rats and mice can be characterized as containing specific chromosome alterations, most frequently translocations and less frequently deletions. These generally cause a change of function of protooncogenes by creating fusion proteins (Lust, 1996; Rabbitts, 1994). There is limited evidence to provide links between human hematopoietic tumors and those in experimental animals. Perhaps the closest link is for human chronic myeloid leukemia (CML) and mouse myelogenous (acute) leukemia (MML). A high proportion of CML cases are characterized by a translocation between chromosomes 9 and 22, involving the *ABL1* protooncogene on chromosome 9 and the *BCR* (breakage cluster region) on chromosome 22. The result of this translocation is the formation of a fusion gene that expresses the *BCR/ABL* chimeric protein, *p190* (Heisterkamp *et al.*, 1990). The great majority of spontaneous and radiation-induced MML is characterized by chromosome 2 deletions (Hayata *et al.*, 1983). Mouse chromosome 2 contains *Abl1* but the deletions do not disrupt this protooncogene (Breckon *et al.*, 1991) and alterations in *Abl1* expression do not appear at first to be involved. More recent data suggest that alteration of a specific tumor-suppressor gene located within a defined region of the chromosome 2 deletion is involved in the development of MML following radiation exposure (Silver *et al.*, 1999). However, this does seem to require further work to determine if the proximity of *Abl1* to the deletions in MML and its involvement in human CML is a coincidence or not. There has been little concerted effort to characterize other rodent hematopoietic tumors, and little information on radiation-induced ones in humans (Section 4).

For oncogene activation, there is a developing database that is indicative of some similarities, at least for the two most commonly mutated protooncogenes, across species for the same tumor type or for related tumor types. This comparison holds for spontaneous tumors and to a more limited extent with ionizing radiation-induced tumors.

5.3.6.2 *Tumor-Suppressor Genes.* The other major class of genes that is mutated in tumors is the tumor-suppressor genes. In contrast to the oncogenes that have a gain of function when mutated (activated), tumor-suppressor genes typically have a loss of function. The function of the normal tumor-suppressor gene is to negatively regulate cell proliferation, and thus their loss or alteration enables the cell to override the normal checks and balances. A list of tumor-suppressor genes involved in tumor formation is presented in Table 5.4.

There are considerable data on mutations in tumor-suppressor genes in a broad range of tumors, although no general review is available. The *TP53* tumor-suppressor gene is generally regarded as being the most frequently mutated in human tumors as evidenced by the review of Harris (1996). In contrast, there is considerably less reported evidence for *Trp53* mutations in rodent tumors, which might, in part, be a consequence of the specific strains used. There are some isolated examples, however, of alterations in *Trp53* reported for rat and mouse tumors, for example, chemically-induced mouse fibrosarcomas (Halevy *et al.*, 1991) and chemically-induced rat zymbal gland tumors (Makino *et al.*, 1992). In generalized terms, it appears that oncogene alterations are predominant in rodent tumors and tumor-suppressor gene alterations in human tumors. Clearly, additional data on the molecular basis of spontaneous rodent tumors is needed to provide a more definitive data base for this general contention.

The identification of a number of human tumor-suppressor genes (Table 5.4) has led to the location of the equivalent genes in mice, and to the subsequent inactivation of these by homologous recombination, knock-out techniques. The *Trp53* homozygous knock-out mouse is viable but develops tumors in a few months, the *Trp53* +/− mice have a high frequency of spontaneous tumors at a number of sites (Donehower *et al.*, 1992) and are sensitive to the induction of tumors by a number of mutagenic chemicals (Harvey *et al.*, 1993). The *Trp53* +/− mice appear to be a reasonable model for humans Li-Fraumeni syndrome, but information on induced tumors is reflective of neither qualitative nor quantitative tumor induction in the general population. The homozygous recessive (*Trp53* −/−) animals are informative for a molecular basis of disease, but not for tumor induction by exogenous agents. A similar argument can be made for other tumor-suppressor gene knock-out mice, with the heterozygous mice being informative for cancer-predisposed subpopulations. To date, information on the spectrum of mouse tumors in heterozygous and homozygous knock-out mice is

TABLE 5.4—*Dominant-acting cancer predisposition genes.*

Syndrome	Chromosome (Gene)	Tumor
Familial adenomatous polyposis	5q (*APC*)	Colorectal adenoma
Breast-ovarian cancer	17q (*BRCA1*)	Breast and ovarian carcinoma
Breast cancer (early onset)	13q (*BRCA2*)	Breast carcinoma
Gorlin Syndrome	9q (*PTC*)	Naevoid BCC
Li-Fraumeni	17p (*TP53*)	Many tumor types
Neurofibromatosis Type 1	17q (*NF1*)	Neurofibrosarcoma, leukemia
Neurofibromatosis Type 2	22q (*NF2*)	CNS tumors
Retinoblastoma	13q (*RB1*)	Retinoblastoma
Von Hippel-Lindau	3p (*VHL*)	Many tumor types, especially kidney
Wilms' tumor	11p (*WT1*)	Kidney lesions in childhood

insufficient to draw a clear conclusion. For example, *Rb* knock-out heterozygous mice do not develop retinoblastoma, although some animals develop pituitary tumors from cells lacking a wild-type allele (Jacks *et al.*, 1992). *Nf1* heterozygous mice appear to be normal up to 10 months of life (Brannan *et al.*, 1994) although a longer period of observation or exposure to radiation or chemicals might reveal a cancer predisposition. The authors offer a number of additional possible explanations but suffice it to indicate that the heterozygous *Nf1* mouse is not a very predictive model of human neurofibromatosis. Additionally, mice heterozygous for the *BRCA1* gene do not develop breast cancer at least up to ~1 y, in contrast to humans where *BRCA1* is involved in early-onset breast cancer (Gowen *et al.*, 1996). These examples serve to illustrate some of the significant differences in tumor development at the molecular level in humans and rodents. The situation with radiation-induced tumors might be simpler or even more complex; the data are not available to make this determination.

Two other animal models of cancer predisposition are available, and these are naturally-occurring mutations as opposed to genetically engineered ones. The Eker rat has an *IAP* insertion into the *Tsc2* tumor-suppressor gene that spontaneously predisposes the heterozygote to develop renal cell carcinomas, and in response to ionizing radiation and kidney carcinogens such as dimethyl nitrosamine (Hino *et al.*, 1993; Walker *et al.*, 1992). It does not appear that this is a specific model for human renal cell carcinoma, since *Tsc2* heterozygotes are not specifically susceptible to renal cell carcinoma development, although kidney lesions do develop in these individuals.

The second model is the Apc^{min} mouse in which *min* is a mutant allele of the murine *Apc* locus (Moser *et al.*, 1995). The *min*/+ heterozygote is predisposed to develop intestinal adenomas, with the number being influenced by modifier loci such as *Mom1* (modifier of *min-1*) (Dietrich *et al.*, 1993). The frequencies of intestinal and mammary tumors can be increased by ENU (Moser *et al.*, 1995). A recent study by Degg *et al.* (2003) showed that the Apc^{min} gene interacts with chromosome 16 segments of BALB/c mice to enhance radiation-induced adenoma development in the upper part of the small intestine. Thus, the Apc^{min} mouse appears to be an informative animal model of human intestinal cancer that involves the *Apc* gene, and could provide useful information on interspecies comparisons for colon cancer development following radiation exposures.

There is very little information available on the involvement of tumor-suppressor gene alterations in radiation-induced tumors. It has been reported that there is a high incidence of point mutations in the *TP53* gene in patients with myelodysplastic syndrome among atomic-bomb survivors (Imamura *et al.*, 2002). However, this relatively new information does not allow for a definitive conclusion about the role of radiation in inducing these mutations. Independent confirmation and further research are needed. There are also some data for lung tumors in uranium miners; these data suggest a variable etiology. Taylor *et al.* (1994) reported an increase in AGG to ATG transversions at codon 249 of the *TP53* gene in 16 of 52 lung tumors from uranium miners. Taylor *et al.* proposed that this specific mutation could be a marker for radiation-induced lung cancer. Bartsch *et al.* (1995) and Vahakangas *et al.* (1992) also studied *TP53* alterations in lung tumors from uranium miners, and Lo *et al.* (1995) studied *TP53* mutations associated with domestic radon. Vahakangas *et al.* (1992) observed *TP53* mutations in lung cancers of 19 patients, but none at codon 249. Lo *et al.* (1995) and Bartsch *et al.* (1995) screened lung tumors specifically for *TP53* codon 249 mutations and found 0/19 and 0/50, respectively. It is

120 / 5. RADIATION EFFECTS

more likely that the codon specific alteration is due to factors other than radon, perhaps mycotoxins (Bartsch *et al.*, 1995). There are no rodent data available for further investigation of this contradiction, or to establish the role of *TP53* mutations in radiation-induced rodent lung cancers.

Additional data are needed before the comparative roles of specific tumor-suppressor gene alterations in human and rodent radiation-induced tumors can be reliably assessed. An additional overriding problem is that there are significant rodent strain differences in spontaneous tumor incidence; these strains reflect genetic predispositions [*e.g.*, Drinkwater *et al.* (1989) and Manenti *et al.* (1994) for the hepatocellular carcinogenesis susceptibility genes]. Thus, qualitative and quantitative comparisons between tumor induction in rodents and humans will not only be reflective of heterogenicity in the human population but also will be very dependent on the mouse or rat strain used.

5.3.7 *Conclusions*

The mechanisms of induction of mutations and chromosomal aberrations by ionizing radiations are similar over a range of mammalian species including humans and involve errors of DNA repair or replication processes. As predicted, this comparison holds for cell killing, given that mutations and chromosomal aberrations are the major perpetrators of cell killing. Thus, qualitative extrapolation across species is reasonable, but quantitatively the situation is less clear-cut. There is relatively little information that provides sensitivity factors for different species or strains, especially *in vivo*. However, what is available suggests that a factor of about two covers the range of sensitivities, and so an extrapolation for radiation-induced tumors from rodents to humans that allowed for this range of two is defensible, but only on the assumption that sensitivity to mutation induction is directly reflective of sensitivity to tumor induction.

The underlying mechanisms of induction of spontaneous and radiation-induced tumors in rodents and humans are similar as regards the involvement of multiple steps, with mutations and chromosome alterations (structural or numerical) at each step. Particular gene alterations involved for a specific tumor tend to be different across species, although the data on radiation-induced tumors are limited. Whether this difference is significant as regards extrapolation is not clear. In addition, there are species-specific host factors that can alter the probabilities of tumor development from initiated cells. These factors need to be investigated further to establish how they influence radiation-induced tumor dose-response models and any extrapolation across species. Additional data on the mechanism of tumor formation will improve the level of confidence in extrapolating from data on rodent tumors to human tumors. At present, it seems reasonable to postulate extrapolating for tumor phenotype (*i.e.* using the concept of acquired characteristics) might prove to be the most rewarding avenue.

6. Extrapolation Models

6.1 Interspecies Correlations of Chemical Toxicities

6.1.1 *Introduction*

A large body of data exists on the development of drugs for which preclinical and clinical toxicity data have been obtained. This Section briefly summarizes some of the experience that has been gained in interspecies correlations between acute reversible toxicity and carcinogenicity of anticancer drugs. This correlation is presented as exemplary of extrapolation models derived from the chemical literature. The relevance to problems of radiation risk extrapolation requires the recognition of three distinctions. First, the acute toxicity of cancer drugs is routinely studied in at least two species of experimental animals, and then promising drugs are given to human subjects under carefully controlled conditions. Second, the dosimetry of the drug is less direct than the dosimetry of external radiation. A variety of pharmacokinetic factors must be considered, and prediction of the concentration of the parent drug and, if necessary, its metabolite(s) at relevant sites of action can be complex. Many pharmacokinetic factors show interspecies variation and variation among members of the human population. Third, the detailed mechanisms of toxicity usually differ between radiation and chemicals. These three distinctions apply to drugs and radiopharmaceuticals. These distinctions also hold for interspecies correlation of acute reversible toxicity and carcinogenicity of radiopharmaceuticals.

6.1.2 *Acute Toxicity*

It is ironic that the best information available on quantitative interspecies toxicology is derived from studies of anticancer drugs. These agents are routinely studied in preclinical systems before they are introduced to the clinic. After a drug shows activity in suitable *in vitro* and *in vivo* screens, decisions must be made concerning the appropriate starting dose, dose schedule, and dose escalation strategy in a clinical setting. A common strategy has been to give the drug to the first group of patients at a dose equal to one-tenth of the dose that is lethal to 10 of mice (LD_{10}), with the dose based on a metric of mg m^{-2} body surface area. If this dose does not produce intolerable toxicity, then other groups of patients are treated at progressively higher doses until a clinical judgment is made that the maximal tolerated dose (MTD) has been reached. This dose then serves as the basis for the treatment of larger groups of patients in Phase II trials designed to explore efficacy. Since a very large number of drugs has been evaluated through this process, there is a substantial database on any correlation between mouse LD_{10} and human MTD.

Freireich *et al.* (1966) confirmed and extended the proposal of Pinkel (1958) that doses of anticancer drugs be based on body surface area. Freireich *et al.* examined available data on the toxicity of 18 drugs in mice, rats, hamsters, monkeys, dogs and humans. These drugs included alkylating agents, antimetabolites, and chemicals exhibiting other mechanisms of action. Doses were based on an LD_{10} in small laboratory animals and on the MTD in monkeys, dogs and humans; these were corrected, if necessary, to a schedule of five daily doses. There was a good correlation between the data from the individual nonhuman species and the human subjects as well as between the weighted estimates derived from all of the preclinical studies taken together and experience in the clinic over a dose range of almost four orders of magnitude. The authors ranked the species in order of decreasing predictive ability as monkey, Swiss mouse, BDF_1 mouse, dog, rat, and hamster.

The consequence of a successful correlation using body surface area in the dose metric is that the anticancer drugs tend to be more toxic to humans than they are to the smaller species on a body weight basis. Since body surface area is approximately proportional to body weight to the 2/3 power, doses in mg kg^{-1} are proportional to doses expressed as mg m^{-2} multiplied by body weight to the –1/3 power. Thus, a cancer drug may be expected, on average, to be about an order of magnitude more toxic to a 60 kg human than to a 20 g mouse [$(60,000/20)^{1/3} = 14$]. A reanalysis of these data using a body-weight scaling power of 0.75, calculated to provide the best fit for acute toxicity data by Travis and White (1988) and supported by Watanabe *et al.* (1992), does not change the magnitude of the essential conclusion [$(60,000/20)^{1/4} = 7$].

The success of an allometric correlation for interspecies toxicology appears to derive primarily from pharmacokinetic considerations. If the total body clearance of a drug is proportional to the body weight to a fractional power, n, while the physiological volume of distribution is proportional to body weight, then dosage on the basis of body weight to the power n will have the following consequences: the peak plasma concentration in the smaller species will be larger by a factor of the body-weight ratio to the $1-n$ power; the plasma half-life in the smaller species will be shorter by the same factor; and the area under the plasma-concentration curve (AUC) will be the same in both species.

Collins *et al.* (1986) proposed that pharmacokinetics be used in the design of Phase I clinical trials of anticancer drugs. The idea was that the appropriate metric for exposure is AUC of the drug. The basic hypothesis was that the mouse AUC at the LD_{10} would be predictive of the human AUC at the MTD. This hypothesis was not expected to be valid if there were major differences in the sensitivity of the target cell or schedule-dependent toxicity. Pharmacokinetically guided Phase I trials would then consist of the following steps:

1. Determine the mouse LD_{10};
2. Measure the mouse AUC at the LD_{10}, which would be the target exposure;
3. Begin testing humans at a dose historically shown to be generally safe (1/10 mouse LD_{10} in mg m^{-2});
4. Measure human AUC at the starting dose; and
5. Choose an escalation strategy based upon how close human AUC is to target AUC.

A review of data for 16 drugs by Collins *et al.* (1990) showed generally excellent agreement between the mouse AUC at the LD_{10} and the human AUC at the clinically determined MTD. Thirteen of the 16 drugs showed AUC_{human}/AUC_{mouse} ratios between 0.6 and 1.3. The maximum discrepancy was observed for fludarabine (ratio = 0.1). This drug requires activation by deoxycytidine kinase, which is found at 10-fold higher levels in human than in mouse bone marrow. A subsequent review by Graham and Workman (1992) was generally supportive of the concept. One (rhizoxin) of eight agents reviewed exhibited considerable disagreement between the expected mouse and human toxicities, while another (iododoxorubicin) emphasized the need to account properly for active metabolites when there are marked interspecies differences in metabolism. Fuse *et al.* (1994) divided drugs according to their mechanism of action: Type 1 for cell-cycle phase-nonspecific agents, and Type 2 for cell-cycle phase-specific agents. A good correlation ($r = 0.90$) was observed between the log of the AUCs at the mouse LD_{10} and human MTD for Type 1 drugs.

While there are good theoretical reasons to believe that the AUC of a direct acting alkylating or platinating agent should correlate with a toxic effect, this metric may not generally be suitable for antimetabolites. Antimetabolites are required to be above an inhibitory level to exert their fundamental biochemical effect, and, once that level is reached, the time of exposure may be the most important factor. This can lead to a schedule dependence, which may be substantial under some circumstances. The most extreme example that has been studied carefully appears to be the folic-acid antagonist methotrexate. Zaharko (1974) observed a single-dose LD_{50} of 350 mg kg^{-1} in mice. Divided doses produced comparable or greater toxicity at lower total doses. The effect on lethality of a continuous infusion of less than 1 µg h^{-1} for 96 h from an implanted device was greater than an LD_{50}. Thus, approximately one of the doses was more lethal when the drug was given as an infusion rather than as a single bolus. The explanation appears to be that the infusion produced plasma levels of the order of ~0.01 µg mL^{-1} (2×10^{-8} M). Survival of the mice was 100 for exposures up to 48 h. While a single dose of 350 mg kg^{-1} produced peak plasma levels of the order of a millimolar, the concentration fell as the drug was eliminated so that it was ~0.01 µg mL^{-1} at 36 h.

6.1.3 Chronic Toxicity

Quantitative predictions of chemical carcinogenicity in humans based on results from animal studies have involved considerable controversy. An early attempt examined six agents (NAS/NRC, 1975). A working hypothesis was suggested that, in the absence of other evidence, the lifetime cancer incidence in humans could be estimated from that in experimental animals if chemical dose is expressed on a basis of mg (chemical dosage) kg^{-1} (body weight) total lifetime exposure. Crouch and Wilson (1979) examined 14 chemicals and concluded that human sensitivity was about equal to or less than fivefold that of mice when the exposure was expressed as mg kg^{-1} d^{-1} averaged over the lifetime of the host. Allen et al. (1988) and Crump et al. (1989) studied 23 chemicals selected because: (1) there was reasonably strong evidence for carcinogenicity in humans or experimental animals and (2) carcinogenic potencies could be quantitated in animals and humans. They determined the TD_{25} (tumorigenic dose in 25 % of the test organisms), the dose rate in mg kg^{-1} d^{-1} that would produce cancer in 25 of subjects who would otherwise not be affected by the cancer. The dose rate was determined as the time average over the age period from 20 to 65 in humans and lifetime subsequent to weaning in the test animals. Allen et al. (1988) and Crump et al. (1989) performed numerous analyses and concluded that all dose metrics examined tended to overestimate human risk except dose rate per unit body weight. Ames et al. (1987) have argued that human cancer risk cannot be estimated quantitatively from animal studies. They have proposed using the TD_{50} (Peto et al., 1984) as a basis for ranking the possible human hazard associated with chemicals. The TD_{50} is the dose rate (mg kg^{-1} d^{-1}) which would halve the percentage of tumor-free animals if given for a standard life span. The concept of the ratio of the dose for human exposure to the dose for rodent potency was introduced as a means of setting priorities.

There are numerous well-known problems that limit confidence in predicting human risk from animal bioassay data for hazardous chemicals. Estimates of human dosimetry in epidemiologic studies are notoriously poor. Experimental animals are exposed to high doses which may cause acute biological effects and which may be associated with pharmacokinetic nonlinearities; human exposure tends to be much lower. The animals are usually exposed to chemicals for a large fraction of their natural life span while human exposure is frequently short term and episodic. Cancer is often mediated by a metabolite or metabolites of the administered chemical, and significant differences can occur in the rates of metabolism among mammalian species. Intrinsic biological sensitivity may differ between rodents and humans.

Some of the problems listed above have been partially addressed. There has been an increased emphasis on the use of biologically based models of carcinogenesis to account for effects such as cell mutation, division, death and differentiation (Moolgavkar and Venzon, 1979; Moolgavkar et al., 1980) and cell proliferation (Cohen and Ellwein, 1990). The role of pharmacokinetics in risk assessment has been discussed extensively (NAS/NRC, 1987). Kaldor et al. (1988) examined the occurrence of second tumors in patients with cancer or polycythemia vera in studies that they judged as carcinogenicity produced by single antineoplastic agents. They compared the incidence of ANLL in humans, including other diagnoses such as preleukemia and myelodysplastic syndrome (cumulative 10 y incidence per gram total dose), with the TD_{50} for all tumors and for hematopoietic tumors in mice and rats. There was an apparently good correlation among four of the five agents for which there were data in common among the species. Anticancer drugs are of particular interest and importance in risk estimation because they are used at doses that are acutely toxic, and the actual amount administered is well known.

Dedrick and Morrison (1992) extended the analysis of Kaldor *et al.* (1988) to incorporate the pharmacokinetics of the three alkylating agents among the five drugs studied. It was reported that carcinogenic potencies agreed reasonably well between rodent and human subjects when the dose metric was the total area under the plasma concentration curve of alkylating species derived from the parent drugs; time-average concentrations greatly underpredicted the human risk from the bioassay results (Figure 6.1). The lack of concordance among tumor types is important to note. The agents tended to be leukemogenic in humans, but the carcinogenic potencies were derived from lymphosarcomas in mice and rats.

A major problem with interspecies correlations including humans is the usual lack of correspondence of the temporal pattern of exposure. Morrison (1987) showed that time-average tissue concentrations provide correct risk estimates when applied to one- and multi-hit models; however, detailed concentration histories are required for analyses based on clonal two- and multi-stage models. Use of time-average concentrations common to most risk estimation procedures produced substantial errors under some circumstances.

6.2 Interspecies Prediction of Summary Measures of Mortality: Relative Risk Models

Dose-response modeling and interspecies extrapolation of health risks are quantitative exercises documented by a vast scientific literature that quite literally spans centuries. Although the diversity of such inquiries is extensive and the borders are vague, these quantitative methods are often divided by whether their motivation is "empirical" or "theoretical." The vagueness of the boundary occurs because theoretical models typically involve parameter estimation for phenomena that are either unobserved or unobservable. Although theoretical models have great potential as hypothesis generating and testing devices, they are complex empirical models when applied to a single data realization. To further complicate matters, there is no statistical theory for discriminating between quantitative models that are not subsets within a larger family of related models (*i.e.*, hierarchical). For this reason, it is not possible to choose statistically between empirical or theoretical models or between parametric, semi-parametric or nonparametric models or between models whose departures from linearity do not arise from the simple addition or multiplication of additional model parameters. Risk models can also be difficult or impossible to compare because of differences in motivation and estimation

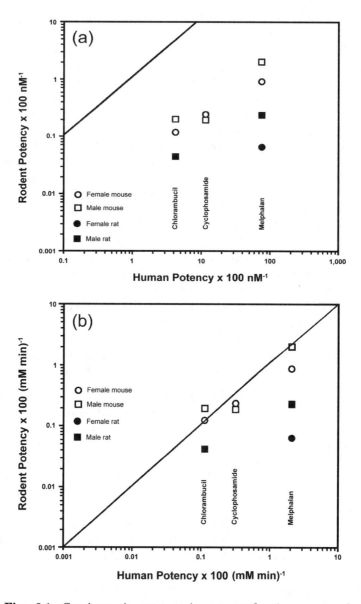

Fig. 6.1. Carcinogenic potency in rats and mice compared with carcinogenic potency in humans. The labels designate the drug that was administered, whereas the potencies are based on exposure to alkylating species derived from the parent drugs. Carcinogenic potency is defined in terms of either the lifetime-average concentration (a) or total exposure (b). The lines of identity between rodent and human potencies are shown for reference (Dedrick and Morrison, 1992).

that arise from the structure of the data being analyzed (*e.g.*, individual death times versus mortality data grouped into age intervals). Despite these problems, however, it is still often possible to make useful inferences about the shape of the dose response, the magnitude of the risk, and the factors that modify the dose response.

Another distinction often used to classify quantitative models applied to failure time data is whether they are absolute or relative risk models. Once again, there are adherents for both types of models. The historical experience of analyses conducted within species as well as comparisons of mortality between species tends to favor relative risk models. Relative risk models are easy to understand and software for these models is widely available. In addition, relative risk models can be easily parameterized to describe patterns of age-specific mortality that cannot be adequately described by a simple multiple of the mortality observed in a baseline group.

Although relative risk models are very useful, absolute risk models also have a role. This is particularly true when the baseline risk is small and when the number of control animals is too small to allow reliable evaluation of the baseline risk and its dependency on age. This was true, for example, in a study evaluating lung cancer risks from radon in rats (Gilbert *et al.*, 1996). Also, absolute risk models are increasingly being used to describe epidemiological data from atomic-bomb survivors and other cohorts (Preston *et al.*, 2002; 2003). Both relative and absolute risk models can generally describe the data, although in some cases the description may be simpler using a relative risk model. Recent analyses of atomic-bomb survivor data suggest that both the relative and the absolute risk depend on attained age.

Finally and maybe most importantly, simple parameterizations of the relative risk model (*via* binary indicator variables) can be used to conceptually and formally link analyses of hazard rates (*e.g.*, Cox model), cumulative survivorship curves [*e.g.*, log-rank test or Kaplan-Meier (1958) analysis], and death rates (*e.g.*, homogeneity chi-square or stratified contingency table analysis). Whether justified theoretically or not, the availability, flexibility and simplicity of the relative risk model makes it the pragmatic method of choice for comparing and predicting injury-related mortality between and within species. Despite this informal endorsement of relative risk models, a reading of the risk literature suggests that models that differ dramatically in structure and motivation often converge to a common interpretation of the data when carefully applied.

6.3 Interspecies Correlations of Radiation Effects

6.3.1 *Introduction*

The need for more detailed knowledge of the specific consequences of radiation exposure of humans became an acute concern during 1941 of the U.S. nuclear program. Initial studies with mice were designed to evaluate the appropriateness of the "tolerance dose" then used for radiation protection. These investigations were started in 1941 to 1943 at the National Cancer Institute under the auspices of the Manhattan Project, U.S. Corps of Engineers (Lorenz *et al.*, 1947; 1954). At the same time, biomedical studies began at a number of universities, such as Chicago, Rochester, Columbia, and also at the newly established atomic-energy facilities.

6.3.2 *Predictions of Radiation-Induced Mortality*

A unique effort began at the University of Chicago's Metallurgical Laboratory Manhattan Project concerning the feasibility of carrying out an extrapolation of radiation effects from experimental animals to humans. Although initiated in the early 1940s, a clear presentation of a specific methodology was first presented by Brues and Sacher (1952) and then further developed by Sacher (1956b; 1966). Sacher's work proposed the use of the Gompertz function (Gompertz, 1825) to define mathematically the age-related patterns of mortality in irradiated and unirradiated populations in terms of the actuarial (Gompertz) equation for the mortality rate, r:

$$r = r_0 \, e^{k_0 t}, \tag{6.1}$$

where in linearized form, r_0 is the intercept and k_0 is the "mortality-rate slope" or regression of age-specific mortality on age, and t is age.

The use of the above equation involved the following four explicit assumptions, for which some experimental evidence then existed (Brues and Sacher, 1952):

1. The Gompertz function for natural aging increases linearly with age.
2. After radiation exposure, the resulting Gompertz function is the sum of contributions due to aging and injury.

6.3 INTERSPECIES CORRELATIONS OF RADIATION EFFECTS / 131

3. A dose of radiation delivered in a short time causes displacement upward without change in slope (after an initial accumulation period, which is neglected here).
4. The Gompertz function for a succession of radiation doses is given by the sum of the effects of the separate doses involved.

The simplest algebraic consequences of these assumptions are that single doses produce dose-dependent displacements of the values of r_0, the intercept, and protracted exposures produce dose-dependent increases in k_0, the slope with no change in the intercept. Obviously, the assumptions could be challenged for a number of biological reasons such as the efficacy of long-term recovery rates, specific tumorigenic responses, and nonadditive interaction with aging processes. It is notable, however, that the third assumption does hold for mice and dogs (Andersen and Rosenblatt, 1969; Sacher, 1956b; 1966). In addition, preliminary long-term survival data on the Japanese survivors in Hiroshima and Nagasaki suggest that the third assumption may hold for humans as well.

The fourth assumption had been periodically challenged on the basis of the "wasted radiation" concept (*e.g.*, Mole, 1955), but that issue has been satisfactorily addressed (Sacher and Grahn, 1964). It should be noted that the assumptions underlying the use of the Gompertz function for the descriptions of the effects of chronic radiation injury in populations have compelling features. The Gompertz function provides mathematically simple and radiobiologically conservative solutions to interspecies extrapolation modeling.

Upon rearrangement of the equation for the 50th percentile of the corresponding cumulative survivorship function, a Gompertz-based estimate of median survival time (MST_0) of mice used as controls can be generated:

$$MST_0 = \left(\frac{1}{k_0}\right) \ln\left[1 + 0.693\left(\frac{k_0}{r_0}\right)\right]. \tag{6.2}$$

Dose-response functions expressed as an *MAS* response variable for irradiated mice were assumed to follow the same relative risk form as the hazard function used to describe mortality in the control population:

$$MAS_D = MAS_0 \, e^{-\beta D}, \tag{6.3}$$

where MAS_D is the MAS value for a group exposed to radiation, MAS_0 is the MAS for the control group, and β is the slope coefficient for the daily dose rate (D). Dose-dependent changes in the Gompertz aging parameter (estimated for each exposure group separately) were also assumed to adhere to the functional form of a relative risk model:

$$k_D = k_0 \, e^{xD}, \tag{6.4}$$

where k_D is the "death-rate slope" for a specific exposure group and x is the coefficient for the daily dose rate D. With the relationship established for "death-rate slopes," MST values for exposure groups (MST_D) were estimated by substituting the k_0 term in Equation 6.2 for the k_D term in Equation 6.4:

$$MST_D = \left(k_0 \, e^{xD^{-1}}\right) \ln\left(1 + 0.693 \, k_0 \, e^{xD/r_0}\right). \tag{6.5}$$

Two additional assumptions were required to establish quantitative linkages between Equations 6.1 through 6.5. First, a relationship between MAS and MST was established by assuming that the irradiated/control ratios for these two endpoints were equivalent:

$$\frac{MAS_D}{MAS_0} = e^{-\beta D} = e^{-xD^{\text{fn}(z)}} = \frac{MST_D}{MST_0}, \tag{6.6}$$

where the power fn(z) was a complex function involving k_0, r_0, x, and D. Second, was the assignment of a theoretical value of 0.831 to the fn(z) supported by an empirical estimate of 0.844 derived from:

$$\ln(MAS_D) = \ln(MAS_0) + \alpha \ln\left(\frac{k_D}{k_0}\right). \tag{6.7}$$

Once Equation 6.6 is established, "all the previous relationships between after-survival, daily-dose, mortality ratio, and death-rate slope fall together" (Grahn, 1970):

$$\frac{MAS_D}{MAS_0} = \left(\frac{k_D}{k_0}\right)^{-0.831} = e^{-\beta D} = e^{xD^{-0.831}}. \tag{6.8}$$

6.3 INTERSPECIES CORRELATIONS OF RADIATION EFFECTS / 133

The relationships described by Equation 6.8 permit the estimation of either the daily dose rate or mortality slope required to reduce survival by 50 %.

Grahn (1970) argued that "surely, we can all agree that all mammalian species have substantially the same form of actuarial experience." From this premise, the ratio of Gompertz aging parameters for mouse and human was presented as a method for converting the radiation risks estimated for mice to those that would be expected for humans. Although expressed more obscurely, it is interesting to note that the same scaling approach was proposed early in this century by Greenwood (1928) as a way to compare the mortality experience of species with dramatically different life spans.

Using mouse data from ANL studies and actuarial data for Swedish males in the period 1956 to 1960, Grahn determined that a scaling factor of 30 was a reasonable estimate for either the ratio of Gompertz slopes or the ratio of Gompertz-derived *MST* values. Two examples were provided as empirical support for this approach to interspecies scaling.

First, it was demonstrated that Swedish males and male BCF_1 control mice had nearly identical death rates for all causes of death or neoplastic deaths when plotted on a time axis derived from the ratio of Gompertz slopes. Next, after the radiation risk coefficients for mice were "humanized" (*i.e.*, multiplied by 30), the relationships described in Equation 6.8 were used to estimate the daily dose required to produce the reductions in life expectancy reported for American radiologists in the 1960s. The resulting daily dose estimate was in agreement with reasonable adherence to exposure standards recommended by NCRP during the 1940s and 1950s. Thus, the radiation protection standards at that time were reasonable.

The approach used to extrapolate radiation-induced carcinogenic risks from mouse strains to humans at Oak Ridge National Laboratory was also based on basic life-table statistics (Storer *et al.*, 1988). Standard actuarial methods for data grouped by age intervals (Elandt-Johnson and Johnson, 1980) were used to estimate the midinterval hazard rate [$\lambda(t)$], rather than making explicit distributional assumptions as was done at ANL. Thus:

$$\lambda(x) = \frac{d(x)}{n(x)}, \quad (6.9)$$

where $\lambda(x)$ is the midinterval estimate of the hazard rate in interval x, $d(x)$ is the number of deaths in the interval, and $n(x)$ represents the mouse days at risk in the interval (*i.e.*, the cumulative sum of days survived by mice within the age interval). The direct method of age-adjustment (Anderson *et al.*, 1980) was used to generate the expected number of mice dying in an interval for each exposure group. In other words, it was assumed that the exposure groups had an identical distribution of days at risk as the control population; the number of neoplastic deaths in an age interval for an exposure group could then be calculated by multiplying the observed $\lambda(x)$ for the exposure group by the days at risk observed for that interval in the control population. The age-adjusted frequencies were then summed over all age intervals to generate the total number of neoplastic deaths expected to occur within an exposure group. For each exposure group (d), an average hazard rate $y(d)$ was generated when the expected total number of neoplastic deaths was divided by the total mouse days at risk observed in the control population:

$$y(d) = \Sigma n_0 \left[\frac{x \lambda_d(x)}{\Sigma n_0(x)} \right], \tag{6.10}$$

where $n_0(x)$ is the mouse days at risk in interval x for the control population, and $\lambda_d(x)$ is the observed hazard rate in interval x for dose group d. To adjust for missing age intervals at the end of the life table in those groups receiving larger exposures, the last age interval used in the calculation of the average hazard rate for all groups was determined by the last interval observed for the shortest lived treatment group.

The $y(d)$ values were then used as the response variable in ordinary least squares statistical analyses to determine whether absolute or relative risk models provided better fits to the dose-response data. A model was parameterized for each sex-strain combination in such a way as to accommodate either type of risk model within a single general equation:

$$y(d) = a + bd, \tag{6.11}$$

where a is the intercept, and b is a slope coefficient that depends on g and h in the form $b = g + ha$. If $h = 0$ in this parameterization, the model becomes an absolute risk model and if $g = 0$, Equation 6.10 then reduces to a relative risk model. When this modelling approach was applied, it was found (Storer *et al.*, 1988) that "the relative risk model is the more preferable model for extrapolating risk of fatal cancers across mouse strains." When relative risks were empirically compared with those observed for the Japanese survivors of Hiroshima and Nagasaki, it was concluded that "relative risks, at least for some tumors, extrapolate directly across mouse strains and from mouse to human with no adjustments required for longevity, body size, metabolic rates, etc."

The ANL approach to extrapolating life-shortening effects (Sacher, 1956b; 1966) and the Oak Ridge National Laboratory approach that involves the direct extrapolation of relative risks across species boundaries share several features. Both approaches to the extrapolation problem depend on synoptic measures (*e.g.*, *MAS*, *MST*, Gompertz aging parameter, age-adjusted average hazard rate) that collapse the entire life table into a single summary statistic. The synoptic measures, in turn, are linked to the dose effect *via* ordinary least squares analyses. In strains with different baseline risks, the relative risks were found to be more comparable than absolute risks. Finally, the dose-response equations used in both approaches have the functional form of relative risk models.

6.3.3 *Example of Interspecies Prediction for Single Exposure*

The method developed at ANL (Brues and Sacher, 1952; Sacher, 1956b; 1966; Sacher and Grahn, 1964) for estimating days of life lost per unit dose (discussed above) was recently examined for its potential relevance to the interspecies prediction of radiation-induced mortality (Carnes *et al.*, 2003). Single exposure data (gamma rays) for the $B6CF_1$ mouse (ANL studies), beagle (Davis study; Andersen and Rosenblatt, 1969), and atomic-bomb survivors were used for the analyses. Intrinsic mortality [defined as deaths that arise primarily from the failure of biological processes that originate within the organisms (Carnes and Olshausky, 1997)] and deaths caused by solid tissue tumors were defined as the two pathology endpoints of interest in these analyses. Death (hazard) rates were assumed to follow a Gompertz distribution:

$$\lambda(t) = \alpha\, e^{\beta t}, \tag{6.12}$$

where $\lambda(t)$ is the hazard rate, t is the failure (or censoring) time, and α and β are the Gompertz parameters. Use of the natural logarithm of the Gompertz hazard function permits the Gompertz parameters to be estimated by linear regression (Gehan and Siddiqui, 1973):

$$\ln[\lambda(t)] = \ln(\alpha) + \beta t, \tag{6.13}$$

where $\ln[\lambda(t)]$ are the natural log of the midinterval estimates of the hazard rates generated from a life table, t is the midpoint of an age interval, the intercept term of the regression equation estimates the log of α, and the slope coefficient provides an estimate of β (with dimension t^{-1} in Equation 6.12).

Next, an analysis of covariance was performed to test the assertion that for single exposures to radiation, the dose groups will share a common slope term (see discussion of the third assumption above):

$$\ln[\lambda(t)] = \ln(\alpha_0) + \Sigma I_i + \beta_0 t + \Sigma \beta_i t I_i, \qquad (6.14)$$

where α_0 and β_0 are the regression terms for the Gompertz parameters associated with the control group, and I_i are binary (0 or 1) indicator (dummy) variables for groups that identify ($I_i = 1$) as a specific dose. The analysis of covariance model described in Equation 6.14 is equivalent to fitting Equation 6.13 to each dose group separately, and the hypothesis that all dose groups share a common slope is tested by determining whether $\beta_i = 0$ for all i other than $i = 0$. If the common slope hypothesis cannot be rejected, then an analysis of covariance model with variable intercepts (i.e., each dose group has its own intercept) and a common slope is fitted to the death rate data:

$$\ln[\lambda(t)] = \ln(\alpha_0) + \Sigma I_i + \beta t. \qquad (6.15)$$

The next step of this procedure is to regress the intercepts $[\ln(\alpha_0) + \Sigma I_i]$ estimated from Equation 6.15 on dose:

$$\ln(\alpha_i) = A + Bd \qquad (6.16)$$

where A is now the intercept, and B is the estimated regression coefficient of dose (probability of death per unit dose). The estimate of days lost per unit dose is then calculated as B/β. The results of these analyses are summarized in Table 6.1.

As predicted (see the third assumption above), the hypothesis could not be rejected that dose groups within strata (defined by species, sex and pathology endpoint) exhibited the same pattern of risk with age (see β_i column in Table 6.1). This means that for single exposures to radiation, every dose group has the same Gompertz rate term (i.e., β in Equation 6.12). The age-specific death rates for

6.3 INTERSPECIES CORRELATIONS OF RADIATION EFFECTS / 137

TABLE 6.1—*Summary of information used to estimate days of life lost per unit dose.*

		Intrinsic Mortality				Solid Tissue Tumor Mortality			
Species	Sex	$\beta_i = 0$ [a] p-value	$\beta (10^{-2})$ [b]	$B (10^{-2})$ [c]	B/β [d]	$\beta_i = 0$ [a] p-value	$\beta (10^{-2})$ [b]	$B (10^{-2})$ [c]	B/β [d]
			(standard errors) [e]				(standard errors) [e]		
Mouse	Male	0.79	0.6270 (0.0250)	0.287 (0.059)	0.46 (0.10)	0.29	0.5260 (0.0349)	0.211 (0.062)	0.40 (0.12)
	Female	0.18	0.6060 (0.0290)	0.268 (0.017)	0.44 (0.03)	0.80	0.5190 (0.0303)	0.375 (0.038)	0.72 (0.08)
Dog	Female	0.10	0.1040 (0.0073)	0.207 (0.031)	1.99 (0.33)	0.12	0.1400 (0.0097)	0.314 (0.058)	2.24 (0.44)
Human	Both	0.19	0.0187 (0.0006)	0.377 (0.0322)	20.16 (1.82)	0.11	0.0203 (0.0009)	0.368 (0.053)	18.13 (2.71)

[a] Significance level (p-value) ($\beta_i = 0$) for slope equality test (Equation 6.14).
[b] Common β slope (Equation 6.15).
[c] Slope coefficient for regression of Log(α) on dose (Equation 6.16), B.
[d] Estimate of days lost per unit dose = B/β.
[e] Standard errors for the parameter estimates are in parentheses. Note: days lost for tumor mortality in male mice increases to 0.56 when dose range restricted to <4 Gy.

the dose groups within a stratum form a visual pattern of parallel lines when plotted on semi-log graph paper. The displacement of these lines from the control (*i.e.*, intercept difference) provides an estimate of radiation-induced damage for the exposed groups. When the intercepts from Equation 6.15 (variable intercept, common slope analysis of covariance) are regressed on dose, the resulting slope coefficients (β in Equation 6.16) are reasonably invariant across species and sex (Table 6.1). Dividing this β term by the common slope generates estimates of the days lost per 10 mGy that range from approximately 0.5 d for the mouse, to 2 d for the dog, and 20 d for humans (Table 6.1). In this metric of radiation injury, the slope term for a species (β) acts like a scaling device to express the days lost per unit dose on an appropriate species time scale.

6.3.4 *Conclusion*

It is important to appreciate the significance of this simple analysis. At least for single exposures, the slope term (β) is statistically invariant across dose groups within a species. This means that β can be estimated from a well chosen control population (*i.e.*, dose data are not needed). Similarly, the relative invariance of the B term across species means that any species with good dose-response data (*e.g.*, laboratory animals) can be used to estimate this parameter. This means that defensible quantitative estimates of radiation injury can be determined for species (*e.g.*, humans) even when the exposure data of interest are either lacking or of poor quality.

6.4 Interspecies Prediction of Age-Specific Mortality

6.4.1 *Introduction*

Since about 1970, estimates of the lifetime effects of radiation exposure have been based upon the accruing data from several long-term studies of irradiated human populations or cohorts [see, for example, NAS/NRC (1972; 1980; 1990) and UNSCEAR (1982; 1986; 1988; 1994; 2000) reports]. These reports reveal an expanding database and increasingly sophisticated methods of analysis to produce estimates of life shortening and of excess risks of cancer mortality. Both absolute and relative risk models have been used, with the latter gaining favor as the data have become more complete. The most recent analyses of atomic-bomb survivor data use both relative and absolute risk models, with allowances for dependencies on age at exposure and attained age. It is increasingly recognized that both models provide useful descriptions of the data.

The major sources of human data are from low-LET irradiations given as single or briefly fractionated exposures of the whole body or significant portions thereof. Data for low-dose low-LET exposures include reports by Cardis *et al.* (1995) and Gilbert (2001). For higher protracted doses and plutonium exposures, the Mayak worker cohort in Russia is discussed by Cardis and Esteve (1992), Gilbert *et al.* (2000), Koshurnikova *et al.* (2000; 2002), Kreisheimer *et al.* (2003), and Shilnikova *et al.* (2003). Data from high-LET radiations and from low-dose rate, low total dose, fractionated/protracted exposures are either limited or unavailable for human populations. Data from animal studies are, therefore, still essential for extrapolation. As both life shortening and all cancer mortality can be used for accurate projections of expectations in human cohorts, high quality animal survival data concurrent with adequate pathology can compensate for the deficiencies in the human database. This is critical for those agencies that need to make estimates of the lifetime responses to high-LET radiations (neutrons and heavy ions) and to low dose, low-dose rate, periodic or continuous exposure to predominately low-LET radiations.

6.4.2 *Background and Justification for Interspecies Predictions*

The risk estimates generated from quantitative models used for the prediction of radiation-induced mortality are often expressed as multiples of those observed in a control population. A control population, however, is composed of individuals that are aging. As a consequence, the mortality risks for exposure to radiation generated by these models must be interpreted relative to the age-specific risks of death associated with growing old, the biological basis of which has been explored by evolutionary biologists [see Carnes and Olshansky (1993) for an overview].

A basic tenet of evolutionary biology is that the force of natural selection alters the genetic composition of a population through the differential reproductive success of individuals. Evolution theory predicts that, starting with the age of sexual maturity, the ability of selection to affect the frequency of competing alleles in a population begins a progressive decline that continues throughout the reproductive period of a species. Thus, diseases that are subject to a genetic influence (referred to as intrinsic diseases) should be influenced by the temporal behavior of natural selection (Medawar, 1952), leading to the prediction that age-specific death rates for intrinsic diseases should rise with advancing age as the influence of natural selection progressively wanes (Carnes *et al.*, 1996).

The gene pool of every sexually reproducing species contains many genes including those that cause intrinsic disease and those that predispose the carriers of those genes to intrinsic disease processes. A shared evolutionary history ensures that many of these intrinsic diseases will be the same in different species. In addition to the heritable disease component, intrinsic diseases also include those thought to arise from genetic damage that is accumulated over time in somatic and germ cells from sources like metabolically derived free radicals (Carnes and Olshansky, 1997). When viewed collectively, the constellation of intrinsic diseases that characterize a species must be unique because every species has a unique genome, despite the fact that there are many genes in common.

The age-related gradient for the effectiveness of selection leads to a natural partitioning of the life span of sexually reproducing species into biologically meaningful age ranges, *i.e.*, a pre-reproductive period where the effectiveness of selection is high, a reproductive period where selection wanes, and a post-reproductive period where selection is either negligible or nonexistent (Carnes *et al.*, 1996). This is the playing field upon which intrinsic causes of death compete for the lives of individuals, *i.e.*, a mortality template for sexually reproducing organisms, whether they arose from nature or were created by geneticists for laboratory experimentation. All that is required is that the organisms have predictable patterns of development and reproduction.

Under this conceptual framework, species are expected to differ in the specifics of what they die from and the time scale over which deaths occur. However, when intrinsic causes of death are considered collectively, the age pattern of death (referred to as an intrinsic mortality signature) is predicted to be similar for different species when viewed on a biologically comparable time scale, *i.e.*, a modern extension of the historical concept of a law of mortality (Carnes *et al.*, 1996; Olshansky and Carnes, 1997). Results consistent with this prediction have been demonstrated for the mouse, dog, and human when the assumption is made that the median age of intrinsic mortality occurs at a biologically comparable age (Figure 6.2).

6.4.3 *Continuous Exposure: Mice to Dogs*

Since unexposed populations of different species appear to share a common intrinsic mortality signature, it was decided to investigate whether exposure to radiation modifies the intrinsic mortality signature of different species in the same way. The data

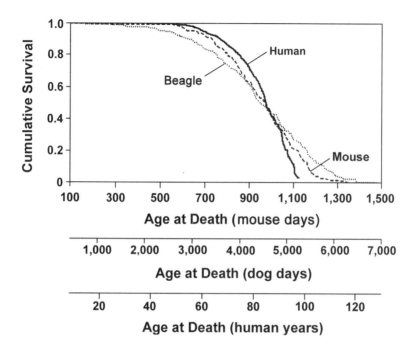

Fig. 6.2. Cumulative survivorship curves of the mouse, beagle and human for intrinsic causes of death (Carnes *et al.*, 1996).

used for this investigation came from mortality data for the B6CF$_1$ (male × female) mouse and the beagle that were compiled in an extensive database at ANL.

Mice entered the study at ~100 d of age and were then exposed to ^{60}Co gamma rays for the duration of life at 11.7, 23.4, 54, or 108 mGy d^{-1} (Grahn *et al.*, 1994). Beagles entered the study at ~400 d of age and were exposed to ^{60}Co gamma rays for the duration of life at 3, 7.5, 18.8, or 37.5 mGy d^{-1} (Carnes and Fritz, 1991). Dose rates for mice and beagles are expressed as average absorbed dose at the midline of the body (Sinclair, 1963). Necropsy and/or pathology information was used to identify extrinsic and intrinsic causes of death. Extrinsic deaths included those caused by infectious and communicable diseases, inflammatory diseases, amyloidosis, environmental stress, accidents, wounds, and iatrogenic events. All other causes of death (including all neoplastic deaths) were defined as intrinsic.

A Gompertz distribution was fit to the control population of the mouse and beagle. The estimated parameters for the Gompertz distribution of each species were used in the cumulative survivorship function to estimate a median age at death from intrinsic causes. The ratio of control medians was used as a scaling constant to adjust for observed differences in the life spans of the mouse and beagle (Carnes et al., 1996).

A proportional relationship between the hazard rate for the control population of $\lambda_0 t$, and the hazard function for each of the irradiated groups of mice, $\lambda(t;z)$, was assumed:

$$\lambda(t;z) = \lambda_0(t)\, e^{f(z)}, \tag{6.17}$$

where t is the age at death (in days), $f(z)$ is a linear function of two covariates (age at entry and dose rate), and $e^{f(z)}$ is the relative risk term of the model (Carnes et al., 1998). Deaths associated with intrinsic causes were considered events and all other deaths were treated as censored observations. The dose-response model for the mouse was then used to generate predicted survival curves for an average age at entry (101 d) and for dose rates that would result in the same accumulated dose at the same relative points of the life span as observed in the beagle (i.e., the dose rates used in the beagle study times the scaling constant). For each of the predicted survival curves, the scaling constant (5.51 dog days per mouse day) was used to convert mouse days to dog days. The only statistical manipulation of mortality data for the beagle involved the generation of Kaplan-Meier (1958) survival curves and their 95 confidence intervals for each of the observed dose-rate groups.

In an analogy to archery, the bull's-eye of the target represents the observed response to radiation by the beagle with the concentric rings around the bull's-eye representing the uncertainty associated with this response. To continue this analogy, the dose-response predictions for the B6CF$_1$ mouse were the arrows and the objective was to determine how close the arrows came to hitting the appropriate bull's-eye of the beagle. As anticipated from previous work (Carnes et al., 1996), the survivorship function predicted for the control population of the B6CF$_1$ mouse fell within the 95 confidence interval of the survivorship function observed for the beagle control group (Figure 6.3). A desired but less anticipated result was that after scaling, the predictions from the mouse dose-response model were also captured by the confidence intervals for the irradiated beagle populations (Figure 6.4), thus reinforcing the utility of the extrapolation model.

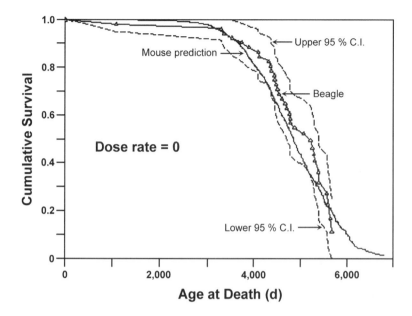

Fig. 6.3. Cumulative survivorship curve and 95 confidence interval (C.I.) for the beagle control group (arrow) and the predicted curve from the dose-response model for the B6CF$_1$ mouse (Carnes et al., 1998).

6.4.4 *Single Exposure: Mice to Dogs and Humans*

Recently, this proportional relationship approach has been applied to the interspecies prediction of radiation-induced mortality for single exposure to gamma rays (Carnes et al., 2003). Once again, B6CF$_1$ mice from the ANL studies were used as the predictor species (Grahn et al., 1995). The mice were irradiated with a single dose of ^{60}Co gamma rays at ~110 d of age; the midline dose ranged from 220 to 7,560 mGy. Mortality data for dogs came from a radiation study involving 354 female beagles conducted at the University of California, Davis between 1952 and 1958 (Andersen and Rosenblatt, 1969). Dogs were irradiated with either single or fractionated exposures to 250 kVp x rays at ages ranging from 10 to 12 months. Average absorbed total doses were either 750 or 2,250 mGy delivered in either one, two or four equal fractions spaced at either 7, 14 or 28 d intervals. For humans, data collected on the survivors of the atomic bomb by the Atomic Bomb Casualty Commission and its successor; the Radiation Effects Research Foundation, were used. This cohort includes ~86,572 survivors for whom dose estimates and mortality information exist (Pierce et al.,

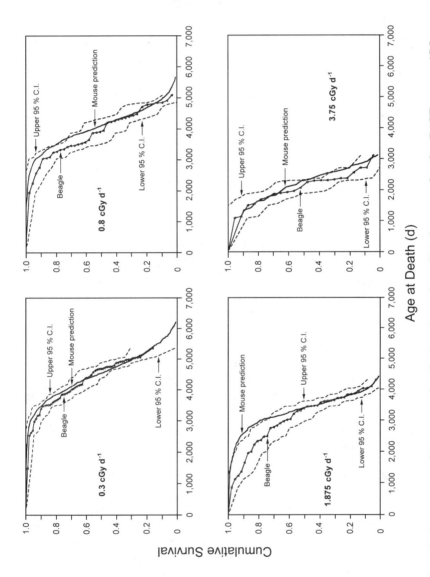

Fig. 6.4. Cumulative survivorship curves for the beagle and predicted curves for the B6CF mouse (C.I. = confidence interval) (scaled to dog days) (Carnes *et al.*, 1998).

1996; Shimizu et al., 1999). Based on their age at the time of exposure, five age cohorts were selected for use in this study: 15 to 20, 20 to 25, 25 to 30, 30 to 35, and 35 to 40 y. Using an RBE of 10 for neutrons, 12 dose groups ranging from 2.5 to 3,500 mSv were used for analysis.

Intrinsic mortality is not a pathology endpoint that has been used in the risk analyses of atomic-bomb survivors. To conform with the traditional endpoints, solid tumors (a major component of intrinsic mortality) were used to compare radiation-induced mortality in the mouse and in atomic-bomb survivors. Solid tumors are classified as tumors other than lymphatic and hematopoietic neoplasms, including tumors of the breast, respiratory system, digestive system, and all others (NAS/NRC, 1990). As in the previous example, the ratio of control medians estimated from the Gompertz distribution was used to adjust for observed species differences in life span. For the beagle, the scaling factor was 4.5 (4,582/1,015) dog days/mouse day for intrinsic mortality, and 3.8 (4,940/1,285) dog days/mouse day for solid tissue tumor mortality. Because the mortality data for humans were aggregated, and because humans (unlike laboratory animals) receive medical care, trial and error were used to identify the scaling factors needed to normalize the mouse and human control populations for these mortality comparisons (32 human days/mouse day for both pathology endpoints).

The scaled survivorship functions for both pathology endpoints predicted from the mouse dose-response model evaluated at zero (control mice) were captured within the confidence intervals of their counterparts in the beagle and human. This result suggests that the age pattern of death for intrinsic and solid tissue tumor mortality occurs at the same relative time points within the respective life spans of these three mammalian species. With only minor exceptions, the three species also exhibited nearly identical age patterns of radiation-induced mortality for the two pathology endpoints. Mouse predictions for all of the beagle dose groups are provided (Figure 6.5). Although mouse predictions exist for every human dose group in Table 6.2, only those falling within the dose range observed for the beagle are presented (Figure 6.6).

In each of the two figures (Figures 6.5 and 6.6), a proportional hazard dose-response model is presented for mice to predict survivorship curves (i.e., age-specific mortality) associated with two different causes of death [intrinsic mortality (A through C) and mortality from solid tissue tumors (D through F) for the control group and two representative dose levels (single dose) for the species being predicted (the beagle dog in Figure 6.5 and the atomic-bomb survivors in Figure 6.6)]. The solid circles (in both

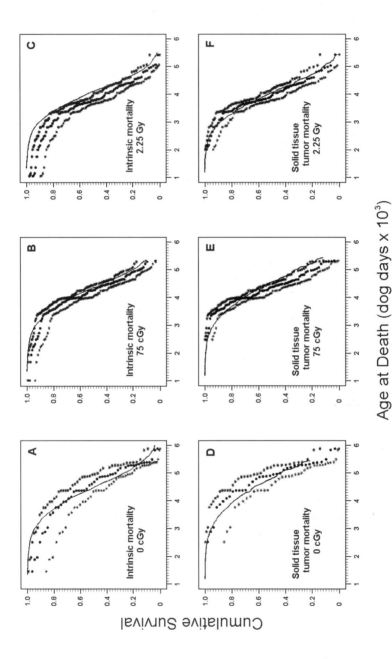

Fig. 6.5. Cumulative survivorship curves (scaled to dog days) for irradiated B6CF$_1$ mice (—) predicted from the dose-response model (Equation 6.10) evaluated at the same total doses received by the beagle (●), and the empirical survival curve [with 95 confidence intervals (∗)] observed for each of the irradiated beagle populations (Carnes *et al.*, 2003).

TABLE 6.2—Summary information for the data used to represent the atomic-bomb survivors.[a]

	Intrinsic Mortality[b]				Solid-Tissue Tumor Mortality[b]				
cGy	T	n	PYR	MaD (d)	cGy	T	n	PYR	MaD (d)
0.1	1,092	862	144,676	31,424	1.0	1,796	520	206,421	35,815
1.0	1,795	1,458	206,926	30,534	3.2	1,055	321	121,784	36,749
3.2	1,055	852	121,974	30,771	7.1	722	211	84,929	35,615
7.1	723	576	85,144	30,709	14	564	182	71,715	34,254
14	560	452	71,769	31,421	32	784	248	88,548	34,133
32	780	634	88,710	30,596	61	268	87	31,683	35,918
60	266	223	31,670	29,924	86	172	55	17,931	34,350
83	174	143	18,016	29,218	123	177	78	17,773	32,325
114	177	149	17,848	29,539	172	91	32	7,705	34,895
154	91	77	7,705	28,341	222	39	15	3,701	33,422
210	66	60	5,538	27,073	274	27	10	18.7	28,494
276	27	23	1,720	24,210	345	27	6	1,720	29,070
312	24	21	1,971	27,806	459	24	10	1,971	29,698

[a]Carnes et al. (2003).
[b]cGy = total dose received.
T = total number of deaths.
n = number of deaths caused by either intrinsic of solid tissue tumor mortality.
PYR = cumulative person years at risk.
MaD = Gompertz median age at death expressed in days.

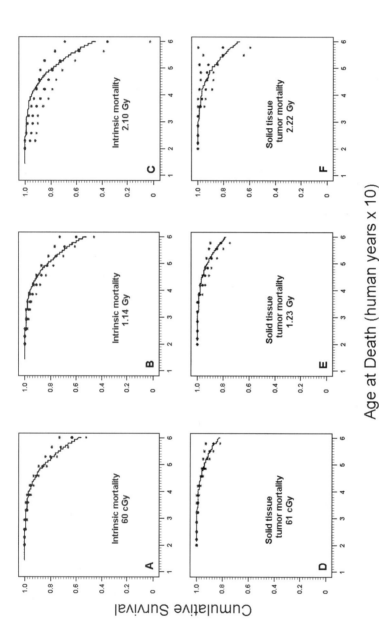

Fig. 6.6. Cumulative survivorship curves (scaled to person years) for irradiated B6CF$_1$ mice (—) predicted from the dose-response model (Equation 6.10) evaluated at the same total doses received by the atomic-bomb survivors (●), and the empirical survival curve [with 95 confidence intervals (✱)] observed for representative groups of the atomic-bomb survivors (Carnes *et al.*, 2003).

plots) are the product-moment estimates (*i.e.*, nonparametric) of the survivorship curve for each target species, and the asterisks are the corresponding 95 confidence interval. The solid line is the prediction generated from the mouse dose-response model. Goodness-of-fit for the predictions can be assessed visually by examining how well the solid line remains within the confidence intervals.

6.4.5 *Conclusion*

The predictions are remarkably good. A simple dose-response model (with no adjustment for host factor differences) for B6CF$_1$ mice exposed to a single dose of gamma rays was able to successfully predict mortality in comparably exposed beagles and humans. These results (as well as the continuous exposure data portrayed in Figure 6.4) provide strong support for the basic thesis of the Report; namely, animal models can be used to make interspecies predictions of radiation-induced mortality, at least for doses below those where host factor differences emerge.

6.5 Extrapolation of Dose-Rate Effectiveness Factors

6.5.1 *Requirements and Limitations*

Since many occupational exposures to ionizing radiations are prolonged exposures, risks for such exposures need to be estimated. Human data for prolonged exposures to almost all types of ionizing radiations are insufficient for risk estimates and can therefore not be used to infer DREFs. It is, therefore, necessary to use experimental animal data to infer DREFs indirectly by extrapolation. Below we give an example of how Bayesian techniques can be used to estimate a DREF for gamma radiation is provided.

Before proceeding with the estimation of the DREF, a brief outline of the Bayesian approach to parameter estimation and characterization of the uncertainty of parameter estimates is given (Bernardo and Smith, 1994).

It is characteristic for Bayesian methods to describe the remaining uncertainty about parameters by probability densities. After updating with observed data incorporated in the likelihood, the so-called posterior density shows the remaining uncertainty about the parameter (Bernardo and Smith, 1994). Subsequently, a model (θ) which contains four parameters, α, β, λ, ρ is considered. The uncertainty about $\theta = (\alpha, \beta, \lambda, \rho)$ is then characterized by $f(\theta|D)$, the joint posterior density of these parameters. In symbols the joint posterior distribution can be written as: $f(\theta|D) \propto f(D|\theta) f(\theta)$ (Bernardo and Smith, 1994). The first factor on the right side is the likelihood and the second the prior density. The likelihood contains the observed data D.

The data used for the extrapolation of a DREF were described previously in detail by Storer et al. (1988) and Ullrich and Storer (1979). Only features of the data essential for an understanding of the subsequent analysis are presented here. At ~10 weeks of age, female BALB/c mice were irradiated with a ^{137}Cs gamma-source at dose rates of 0.4 Gy min^{-1} or 0.083 Gy d^{-1}. A control group of about the same age was sham irradiated.

Among parametric survival models, the Weibull distribution is the typical choice (Hoel, 1972; Kalbfleisch and Prentice, 1980; Kleinbaum, 2005). Its hazard function for the two parameter version (scale and shape) is simply $\lambda(t)^\alpha$ where t represents survival time. To include dose D in the model, a slope parameter β is incorporated with the hazard function becoming:

$$(\lambda t)^\alpha e^{\rho \beta D}, \tag{6.18}$$

with ρ representing the DREF. This is the usual Cox PHM (Kalbfleisch and Prentice, 1980; Kleinbaum, 2005) for a covariate such as dose which is treated here in a linear manner in the model. The DREF term ρ multiplies the dose D so that, for example, a DREF of two implies the need for twice the dose to obtain the same risk as with an acute exposure.

The probability function is then:

$$f(t|\alpha,\beta,\lambda,\rho) = \alpha \lambda^\alpha t^{\alpha-1} e^{\rho \beta D} e^{-(\lambda t)^\alpha e^{\rho \beta D}}, \tag{6.19}$$

$f(t|\alpha, \beta, \lambda, \rho)$ is the density function for a relative risk model which has a background rate ($D = 0$) of a Weibull model and DREF (ρ in the relative risk term e^D). Since the advent of the Cox or PHM, this exponential form of the relative risk model has been popular in the literature. However, a Taylor series expansion would show that $e^D = 1 + \rho \beta D$ for $\rho \beta D \ll 1$. Therefore, this is a linear function. The adequacy of the model was checked by comparing the cumulative hazard of the model obtained by inserting the maximum likelihood estimates for the parameters with the cumulative hazard resulting from the nonparametric estimation procedure of Aalen (1976). The comparison is shown in Figure 6.7. The smooth solid line shows

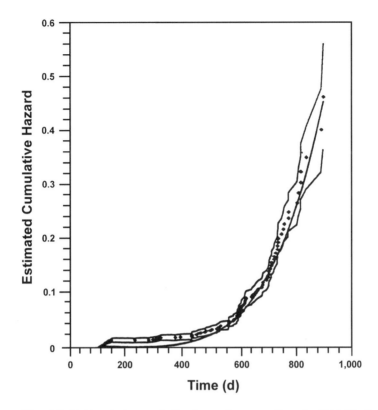

Fig. 6.7. Estimated cumulative hazard versus time for BALB/c female mice with breast cancer (2 Gy at 0.4 Gy min^{-1}). The smooth solid line shows the cumulative hazard of the model using the maximum likelihood estimates of the parameters. The data points are the nonparametric estimates of the cumulative hazard, and the upper and lower lines with discontinuous slopes connect the points one standard deviation above and below the nonparametric estimates (Arnish, 1994; Arnish and Groer, 2000).

the cumulative hazard of the model using the maximum likelihood estimates of the parameters. The data points are the nonparametric estimates of the cumulative hazard, and the upper and lower lines with discontinuous slopes connect the points one standard deviation above and below the nonparametric estimates.

The analysis with this model was performed for several endpoints (Arnish, 1994; Arnish and Groer, 2000). The results of the analysis for mammary tumors are illustrated by Figures 6.8 and 6.9. Each figure shows the marginal posterior density for one of the parameters. These densities are obtained by integrating over all

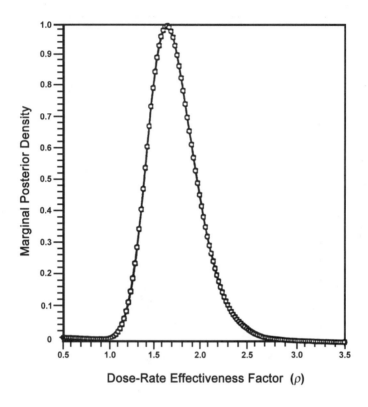

Fig. 6.8. Marginal posterior density for ρ for BALB/c female mice with breast cancer (Arnish, 1994; Arnish and Groer, 2000).

the other parameters. The posterior density of ρ (Figure 6.8) shows that for this endpoint the DREF (ρ) is concentrated around values greater than one. This indicates that for the given model a dose from ^{137}Cs gamma radiation delivered at the higher-dose rate (0.4 Gy min^{-1}) is more effective at inducing breast cancers than the same dose administered at the lower-dose rate of 0.083 Gy d^{-1}, in keeping with the precept that lower-dose rates allow for repair of damaged sites that otherwise would be available for interaction were the dose rate high. With high-dose rates the damaged sites would all be relatively simultaneously present and thus available for interaction.

When lung cancer was chosen as the endpoint of interest the mass of ρ was concentrated at values less than one indicating greater carcinogenic effectiveness of doses delivered at the lower-

6.5 EXTRAPOLATION OF DOSE-RATE EFFECTIVENESS FACTORS / 153

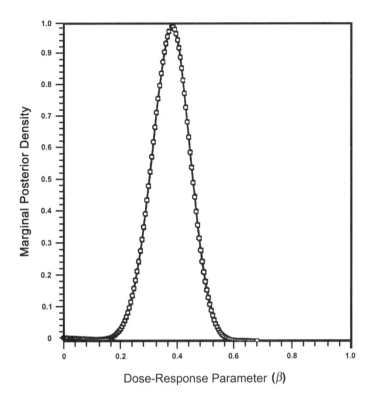

Fig. 6.9. Marginal posterior density for β for BALB/c female mice with breast cancer (Arnish, 1994; Arnish and Groer, 2000). The ordinate for each plot was calculated and connected by the graphing software. β is the "dose-response" parameter for the lower-dose-rate exposure of 0.083 Gy d^{-1}.

dose rate. Figure 6.10 shows the corresponding density for ρ. It is interesting to note that this is an example of an estimated DREF being less than one for low-LET radiation. These factors show that without extensive analysis of animal data, general statements about dose-rate effects in humans should not be made.

To complete the extrapolation of breast cancer risk to humans a density for β_H derived from data on a human population exposed to high-dose-rate gamma radiation is needed. The parameter β_H is the dose-response term for the higher-dose-rate exposure of 0.04 Gy min^{-1}. A density based on the analysis of breast cancer in the atomic-bomb survivor data as presented by NAS/NRC (1990) was adopted. The density of β_H is normal (0.87, 0.013) and is shown in Figure 6.11. Under the assumption of certainty of relevance of ρ

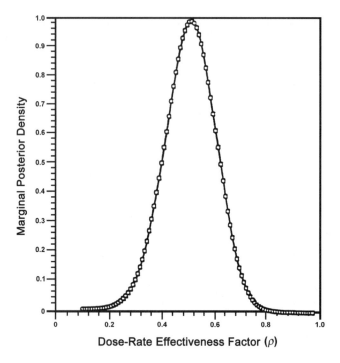

Fig. 6.10. Marginal posterior density for ρ for BALB/c female mice with lung tumors.

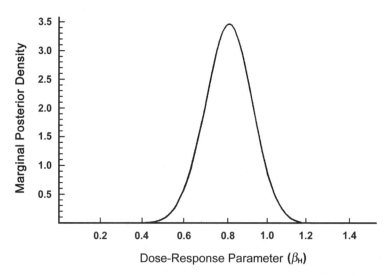

Fig. 6.11. Marginal posterior density for β_H for female atomic-bomb survivors with breast cancer.

for the extrapolation from mice to humans, the dose-response parameter (β_L) for exposure to gamma radiation at the lower-dose rate is: $\beta_L = \beta_H/\rho$. The density for the extrapolated dose-response parameter (β_L) is shown in Figure 6.12. It was derived by using normal densities for β_H normal (0.87, 0.013) and ρ normal (1.6, 0.04). (The second normal density is an approximation for the exact density for ρ shown in Figure 6.10.) The density for β_L characterizes the remaining uncertainty about the dose-response parameter (β_L) for exposure to gamma radiation at the lower-dose rate. This simple example demonstrates that under the assumption of relevance made above, the density for the low-dose-rate response parameter β_L in humans can be estimated by extrapolation using data on mice. This example shows that the probability density for dose-rate effects parameters in humans can be obtained *via* extrapolation from animal data if one is prepared to make certain judgments of relevance. More similar simple analyses of dose-rate effect data in animals and more complex meta-analyses like the one referenced in the next paragraph are needed before general, sweeping statements about dose-rate effects in humans can be made.

A methodologically related but more involved example for extrapolation of risk estimates for plutonium from several animal studies to humans was described by DuMouchel and Groer (1989).

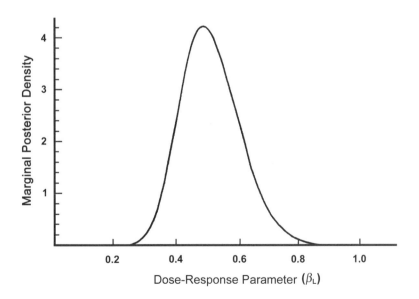

Fig. 6.12. Marginal posterior density for β_L for females with breast cancer after low-dose-rate exposure.

156 / 6. EXTRAPOLATION MODELS

This more complicated approach could also be employed with modifications for the extrapolation of DREFs, provided more relevant animal data on dose-rate effects were available for the analysis.

6.5.2 *Conclusion*

Whenever data for irradiated humans are available, investigators use these data to make their predictions of radiation-induced risk for humans. The data for exposed populations of laboratory animals are almost exclusively used to generate estimates of RBE and DREF values. The successful interspecies predictions presented in this Section should: (1) demonstrate the relevance of the data derived from animal studies for the prediction of human health effects, (2) enhance confidence in the accuracy of RBE and DREF values estimated from animal data for exposure conditions where human data do exist, (3) indicate that RBE and DREF values can be estimated from animal data where human data do not exist, and (4) suggest that mortality data for laboratory animals can be used to predict age-specific radiation-induced risks for humans for general endpoints like life shortening, all cancers, and selected subsets of cancers involving homologous tissues.

6.6 Extrapolation of Results for Internally-Deposited Radionuclides from Laboratory Animals to Humans

There is a continuing need to determine and understand the long-term health risks of internally-deposited radionuclides in persons exposed medically, or occupationally, or from radionuclides in the environment. A full understanding of these health risks, particularly for exposures involving low doses and dose rates, requires in-depth knowledge of both the dosimetry of a given exposure and the resulting long-term biological effects.

Long-term studies of human populations that have dose-response relationships built on both good dosimetry and proper follow-up of long-term biological effects include populations of human subjects that were exposed medically, occupationally or environmentally to radionuclides such as ^{222}Rn, ^{224}Ra, ^{226,228}Ra, ^{232}Th, ^{131}I, ^{239}Pu and their decay products. Results from many of these studies have provided our primary historical sources of direct knowledge on the health risks of chronic alpha irradiation on the lung, liver and skeleton, essential segments of our radiation protection practices for internally-deposited radionuclides.

However, all of the information needed to understand the importance of many dose- and effect-modifying factors such as the elemental characteristics, the physical and chemical forms at exposure, route of exposure, nonuniformity of dose in various tissues and organs, LET, age, gender, etc. cannot be obtained. For this reason, a broad range of complementary studies has been conducted in various species of laboratory animals to understand the similarities and differences associated with the internal deposition and retention of other radionuclides and other radionuclide forms taken into the body by various routes of exposure. *In vitro* systems have also played important roles. The history of research on the dosimetry and health effects of internally-deposited radionuclides in laboratory animals was covered in great detail by Stannard (1988). Other major overview books and reports include Boecker *et al.* (1994), IARC (2001), NAS/NRC (1988), Thompson (1989), and Thompson and Mahaffey (1986). A major use of studies in laboratory animals has been to compare results among various species including humans when available and to extend this information to other interspecies comparisons for which human data are not available. Strategies for some of these comparisons are given as examples in the following sections.

Because the long-term health effects seen in these human populations have occurred primarily in the lung, liver and skeleton, attention in this Section is directed primarily to examples based on these organs. Even with this focus on the lung, liver and skeleton, the range and magnitude of studies conducted in laboratory animals are substantially more than can be covered in a report of this type. Therefore, much of the focus of this Section is directed to results from life-span studies in rodents and beagle dogs because of the interest in stochastic effects, especially cancer. Many other valuable studies have been conducted using other protocols and other species that are not covered here because of space constraints.

Before discussing these studies, some of the aspects of the radiation dose received from internally-deposited radionuclides will be discussed. Of particular interest are the temporal and spatial patterns of dose delivery and the LET of the emitted radiations compared with those for doses received from external sources of penetrating radiation.

6.6.1 *Temporal Pattern of Delivery of Radiation Dose*

The radiation dose from internally-deposited radionuclides comes from the decay of a radionuclide in an organ or tissue. The original deposition of the radionuclide in the tissue or organ depends upon many different factors including the route of entry into the body, *e.g.*, inhalation, ingestion, wounds or dermal exposure. The radionuclide may be deposited directly in the organ or it may be translocated from another site in the body. The temporal pattern of dose depends upon the amount of the radionuclide in the organ, the physical and biological half-lives of the parent radionuclide and its radioactive progeny in the organ, and the elemental chemical properties. These temporal patterns of dose delivery are characterized by a constantly changing dose-rate pattern. In many cases after a single exposure to a long-lived radionuclide, this leads to a pattern where the dose delivery is protracted over a period of many years with a changing dose-rate pattern reflecting the uptake or in-growth of a radionuclide followed by a subsequent decrease due to losses by physical and biological processes over much of that time period. This contrasts sharply with many of the patterns of external delivery of the dose where the dose was delivered essentially instantaneously, as in the case of the Japanese atomic-bomb survivors, or over a relatively short period of time, as occurs with external medical irradiation patterns.

6.6.2 *Spatial Pattern of Delivery of Dose*

After a radionuclide intake, the resulting internal distribution can range from very nonuniform, with deposition and retention occurring primarily in one or a few major organs or tissues, to a broader, more uniform distribution throughout the body depending on the physical and chemical properties of the exposure radionuclide. Two classic examples of a very nonuniform distribution produced by elemental chemical properties are the distributions of ^{131}I, predominantly in the thyroid, and ^{90}Sr, predominantly in the skeleton, after intakes of their soluble forms. The chemical form of the exposure radionuclide can also play a big role in determining the eventual distribution within the body. For example, if exposure is by inhalation to insoluble particles of ^{239}PuO$_2$, the ^{239}Pu will primarily irradiate the lung, and only small amounts will dissolve slowly from the particles and translocate to other organs, primarily the liver and skeleton. However, if the inhalation exposure is to the more soluble form ^{239}Pu(NO$_3$)$_4$, then a much larger amount will dissolve and translocate more quickly from the lung to the liver and skeleton.

These varying degrees of nonuniformity can be contrasted with the patterns seen at the other end of the distribution spectrum, where internally-deposited soluble forms of ^{137}Cs and particularly ^{3}H are distributed rather uniformly throughout the body with a spatial pattern of dose delivery similar to that for external, whole-body, penetrating radiation. A considerable amount of information is available on the distribution of dose among different body organs and tissues in persons having internal intakes of various radionuclides in various forms. ICRP and the Medical Internal Radiation Dose Committee of the Society of Nuclear Medicine have been very active in formulating biokinetic and dosimetric models to account for these differences in uniformity of distribution and irradiation within the body (Howell *et al.*, 1999; ICRP, 1989; 1993; 1995a; 1995b; ICRU, 2002).

6.6.3 *Linear-Energy Transfer (Radiation Quality)*

Linear-energy transfer (LET) of radiation emitted from internally-deposited radionuclides is important primarily because high-LET radiation from the decay of alpha-emitting radionuclides is much more effective in producing biological effects than low-LET radiation. Alpha emitters deposit their energy within 30 to 70 µm from the site of decay in unit-density tissue. This characteristic of alpha-emitting radionuclides leads to unique spatial patterns of dose delivery that have important implications for tumor induction. The two most important examples of this are for alpha-emitting radionuclides that are deposited primarily on bone surfaces, such as ^{239}Pu, and for the radon progeny that decay primarily in the cells of the conducting airways of the lung.

In the case of the alpha-emitting radionuclides on bone surfaces, the sensitive cells for induction of bone cancer are in the endosteal cell layer next to the bone surfaces. This increases the dose delivered by these surface-seeking radionuclides to the underlying endosteal cells compared to radionuclides that are deposited more diffusely throughout the volume of the bone. The radioactive progeny of radon are more likely to decay in the airways of the lung because of their short physical half-lives. The sensitive target cells for cancer induction in people are located in the thin epithelial lining layers of cells in these airways. In both of these examples, it is the short path length of the alpha radiation with target cells within its LET range in a special tissue geometry that leads to unique situations. In most other examples, the alpha-emitting radionuclide has a simpler geometry of random spatial distribution in relation to target cells that sometimes permits the use of an average organ dose.

160 / 6. EXTRAPOLATION MODELS

Because of the differences between external and internal exposure, it has long been recognized that there may be many difficulties in predicting the health effects of internally-deposited radionuclides from external exposure data. The few examples available of human exposure to internally-deposited radionuclides indicate it is prudent to be cautious in predicting the health effects of internal emitters from data on external exposures. For example, people exposed to $^{226, 228}$Ra have long been known to develop bone cancer because of the deposition of radium primarily in the skeleton. The types of bone tumors observed from exposure to $^{226, 228}$Ra would not be predicted from the effects observed in the atomic-bomb survivors. Studies of health effects from internally-deposited radionuclides in laboratory animals provide a means for characterizing the health effects of radiation dose patterns for which data in humans do not exist, but for which it is necessary to limit human exposures.

To evaluate the effectiveness of laboratory animal studies in providing adequate limits for human health protection from internally-deposited radionuclides, it is necessary to compare the results in those few cases where human effects data exist to results in laboratory animal studies. There are many ways in which such a comparison could be carried out. For our purposes, the question is whether the same types of health effects, primarily tumors for a particular organ, are observed in laboratory animal studies and in human epidemiological studies. Asking the questions in this manner will ignore many of the more subtle but important differences between tumors in laboratory animals and people. However, asking the question in this manner does allow for the fact that the main consideration is the protection of human health.

6.6.4 *Internally-Deposited Radionuclides for Which Human and Laboratory Animal Data are Available*

6.6.4.1 *Radium-226, 228.* The primary studies of the health effects of $^{226, 228}$Ra in humans involve the ingestion of radium by dial painters. The radium dial painters were largely young women who used luminous paints containing ^{226}Ra and sometimes also ^{228}Ra. As a consequence of tipping the brush on their lips, the workers ingested significant quantities of ^{226,228}Ra during the early part of the 20th century. Because radium is a metabolic analog of calcium it is deposited in the skeleton. The ^{226,228}Ra retained in the skeleton serves as a source of alpha radiation of bone and nearby tissues at a dose rate that decreases slowly with time (Evans, 1974; Fry, 1998; Rowland, 1994; 1995; Thomas, 1995).

The very long-term follow-up of these workers showed that the biological effects were primarily bone sarcomas and "head" carcinomas, carcinomas of the paranasal sinuses and mastoid processes. Another important feature of these studies was the determination that none of these tumors were seen in subjects with average cumulative skeletal doses below ~10 Gy. Based on these results, Evans proposed a practical threshold for malignancy induction of 3.7 kBq (0.1 µCi) because the latent period associated with intake levels lower than this amount might extend beyond a normal human life span of 70 y (Evans, 1974).

Life-span studies in beagle dogs were conducted at the University of Utah using a single intravenous injection of ^{226}Ra at 18 months of age and at the University of California at Davis using eight semi-monthly intravenous injections of ^{226}Ra starting at 14 months of age (Thompson, 1989). These studies, which parallel human exposure to ^{226,228}Ra, can be compared directly because of the similarity of the dose patterns in the dogs and humans.

Bone tumors are the primary finding in humans and dogs after exposure to ^{226,228}Ra. The dog is much more sensitive to the induction of bone tumors with about 8 to 11 times the rate of bone tumor induction per gray of human exposure based upon a Bayesian analysis (NAS/NRC, 1988b). A slight excess of breast tumors has also been observed in humans and dogs, however, it is uncertain how to interpret these results (Bruenger *et al.*, 1994). Radium does not accumulate in mammary tissue in either species so that the total radiation dose from ^{226,228}Ra is low. In humans, perhaps half of the dose response can be explained by external radiation from the radium solutions. Sinus and mastoid cancers have been observed in humans but not in beagles. These types of cancers occur less than half as frequently as bone cancers. Part of the reason for the difference between species is that the structure of the sinus and mastoids are different in humans and beagles. Radon and its daughter products do not accumulate in the beagle to the same extent as in humans, greatly lowering the resulting dose in beagles.

6.6.4.2 *Radium-224.* The radionuclide ^{224}Ra is in the ^{232}Th decay chain and is the direct progeny of ^{228}Th, an alpha emitter with a radioactive half-life of 1.91 y. The ^{224}Ra, in turn, decays with a radioactive half-life of 3.66 d by emitting alpha particles. Internal exposure to ^{224}Ra occurred as a result of it being injected as a medically therapeutic agent in Europe as a treatment for ankylosing spondylitis or skeletal tuberculosis. Bone tumors have been one of the main health effects seen in a group of patients that have been followed after their injections with relatively high levels of ^{224}Ra as children or adults in Germany between 1944 and 1951. A number of reports on studies of these populations have been published including Mays et al. (1986b; 1989), Nekolla et al. (1999), Spiess (1969; 1995), and Wick et al. (1999).

These populations are listed separately from those exposed to ^{226,228}Ra because the dose patterns in the skeleton are different for ^{224}Ra and produce a different dose-response pattern. The 3.66 d half-life of ^{224}Ra, which is much shorter than that for ^{228}Ra of 5.75 y and ^{226}Ra of 1,620 y, means that the same total dose is delivered to the bone over a much shorter time period. Also, the dose is delivered to different structures of the bone because of the shorter half-life of ^{224}Ra. The longer half-life of ^{226,228}Ra allows time for the majority of it to be distributed into the volume of the bone mineral before it decays. Conversely, the majority of the ^{224}Ra decays while it has been deposited on the surface of the bone and before it has had time to be redistributed into the mineral volume of the bone. The most sensitive tissue of the bone for cancer induction is the lining on the endosteal bone surfaces. The combination of deposition pattern and sensitivity for cancer induction leads to ^{224}Ra being much more effective in causing bone cancer than ^{226,228}Ra.

The people injected with ^{224}Ra were given different numbers of injections, ranging from a few to several hundred. Multiple injections had the effect of protracting the radiation dose over a period of many years thus simulating a radionuclide with a much longer half-life. These results provide a means of evaluating the effects of dose protraction for a radionuclide on the surface of the bone.

Comparable to the studies of injected ^{224}Ra in people is a series of studies in beagle dogs that were begun at the University of Utah and finished at the Lovelace Respiratory Research Institute (Muggenburg et al., 1996b). These studies included dogs that received either 1, 10 or 50 intravenous injections of ^{224}Ra, given at the rate of one injection per week. This schedule protracted the dose in the dogs with 50 injections over ~1 y and provided a comparison of the effect of dose protraction with the study in humans.

As was the case for ^{226,228}Ra, bone cancer was the primary effect seen in humans and beagles from exposure to ^{224}Ra. Dose protraction was also more effective in both species with doses protracted for 1 y (50 injections), with a resulting increase in the rate of bone tumor induction by about 90 times that compared with the same dose delivered over one week (one injection). Similar increases in the rate of bone tumor occurrence were seen in humans when the dose was protracted. Again, the dog was much more sensitive to the induction of bone tumors than humans. Tumors arising from the nasal epithelium were related to the total amount of injected ^{224}Ra in the dogs but similar tumors have not been observed in humans. These tumors arose in the dog from different tissues than did the sinus and mastoid tumors in humans exposed to ^{226,228}Ra, and those tissues are probably irradiated in a different manner. The nasal tumors are much less common than the bone tumors and no effect of dose protraction was found. There was also an excess of breast tumors related to the amount of ^{224}Ra injected in female beagles. Breast tumors have not been demonstrated in women injected with ^{224}Ra. The dose response for breast tumors is difficult to explain in terms of a direct irradiation of mammary tissue because ^{224}Ra does not deposit in mammary tissue.

6.6.4.3 *Thorotrast® (^{232}Th).* Thorotrast® is a colloidal suspension of thorium compounds that was used as an intravascular contrast agent in medical radiography for imaging cerebral and limb blood vessels. After intravenous injection, particles in the colloidal suspension were deposited primarily in the reticuloendothelial system of the liver, spleen and bone marrow where they aggregated in clumps up to 100 µm across that were relatively insoluble. The primary source of the resulting radiation dose was the decay of the ^{232}Th and its long chain of progeny radionuclides. Epidemiological studies have been carried out in Germany, Portugal, Denmark and Japan of populations that were injected with Thorotrast® between the late 1920s and 1955. The results of these studies have been summarized in a number of reports including Gossner *et al.* (1986), Machinami *et al.* (1999), NAS/NRC (1988), NCRP (2001a), and Taylor *et al.* (1989).

Malignancies in the liver and hematopoietic tissues have been the most prevalent long-term biological effects observed in these populations. An important question relating to the general usefulness of these data for estimating liver cancer risks for other alpha-emitting radionuclides deposited in the liver is whether the observed effects were due, in part, to some "foreign body" effect from the several grams of thorium injected in addition to the alpha radiation. A number of studies have been conducted in laboratory animals to examine this question and other related questions in greater detail.

Studies by Wesch *et al.* (1983) involved rats injected with synthetic Thorotrast® to which more radioactivity (^{230}Th) had been added. Others reported by Dalheimer *et al.* (1986) and Wesch *et al.* (1983), involved injection of nonradioactive particles (Zirconotrast,

ZrO$_2$ to which ^{228}Th and ^{230}Th were added). Although some shortening of tumor appearance times may have occurred, the radioactivity appeared responsible for most, if not all, of the liver cancers induced.

Another very different approach to addressing this question was used by Spiethoff *et al.* (1992). They conducted life-span studies of rats given injections of nonradioactive Zirconotrast particles and whose livers were then irradiated with a collimated beam of 14 MeV neutrons from a ^{252}Cf source. The risk coefficient for liver cancer from the neutron irradiation was the same with or without the presence of Zirconotrast particles.

Using another approach, Brooks *et al.* (1986) and Guilmette *et al.* (1989) studied the production of chromosome aberrations in the livers of Chinese hamsters that had been injected with Thorotrast® or soluble ^{239}Pu citrate. In spite of the differences in microscopic distribution of these two forms, similar effectiveness per gray was noted for both, another indication that the foreign-body effect was not an important contributor to the risk of liver cancer from Thorotrast®. Taylor *et al.* (1993) compared the carcinogenic response of the livers of grasshopper mice to injected Thorotrast®, ^{241}Am or ^{239}Pu and found similar risk coefficients for all three radionuclides.

The observation of leukemia in some members of the Thorotrast®-exposed populations is a rather unique finding. Although leukemia has been a prominent biological effect produced by high exposures to external radiation sources, the same is not true, in general, for chronic irradiation from internally-deposited radionuclides. UNSCEAR (2000) noted that human epidemiological data available for internal exposures to low-LET radiation such as the Chernobyl data do not indicate elevated risks of leukemia. For radionuclides emitting high-LET radiation, leukemia risks have only been observed for high-level exposures to Thorotrast®, and to a lesser extent to ^{224}Ra, with little or no evidence for a risk of leukemia following exposure to ^{226,228}Ra or radon and its progeny.

6.6.4.4 *Radon and Radon Progeny.* This is perhaps the best known example of an internally-deposited radionuclide because of the concern over its potential to cause a significant number of cancers in homeowners. Although concern over exposure to radon has more recently focused on exposures in the home, the radioactive progeny of radon has been an occupational problem of substantial magnitude in causing lung cancer in many populations of miners. Historically, the problem was recognized many years ago in the miners in the Erz Mountains of Central Europe who had high incidences of lung cancer and other respiratory diseases. More recently, the mining of uranium and other metals has led to exposures to radon and its radioactive progeny of some miners in North America, Europe, and China.

Many of these populations have been carefully followed in a number of epidemiological studies. These effects have been well documented, reviewed and analyzed in a number of reports including publications from NAS/NRC (1988; 1991), the National Cancer Institute (Lubin *et al.*, 1994; 1995), UNSCEAR (2000), and the U.S. Environmental Protection Agency (EPA, 2003; Pawel and Puskin, 2004).

The exposures of people differed in factors such as the time period over which the dose was accumulated, the dose rate, and the fraction of the radioactive progeny attached to airborne particles. Studies on radon and its progeny use special terms to describe the concentration in the exposure atmosphere and the cumulative exposure that account for different mixtures of radon progeny that may be present in the exposure atmosphere. The term working level (WL) is used to describe the exposure concentration; WL is defined as "any combination of short-lived radon daughters in 1 L of air that will result in the ultimate emission of 1.3×10^5 MeV of potential alpha energy." In a similar way, the term working-level month (WLM) is a measure of cumulative exposure defined as the "exposure resulting from inhalation of air with a concentration of one working level of radon daughters for 170 working hours" (NAS/NRC, 1988).

Factors such as these have been systematically studied in laboratory animals to better understand how to model the dose-response relationships and many of them are summarized in NAS/NRC (1988a) and other more recent reports (Cross and Monchaux, 1999; Heidenreich *et al.*, 2004). Cross and his colleagues at the Pacific Northwest National Laboratory (PNL) carried out many of these studies primarily using rats as the laboratory animal. In these studies, they varied the exposure concentration (10 to 1,000 WL), the total exposure (20 to 10,240 WLM) and the fraction of radioactive progeny attached to airborne particles. Similar experiments were carried out in a different strain of rat in France at the Compagnie Generale des Matieres Nucleaires (COGEMA) laboratory. Each of these laboratories has used 5,000 or more rats in their studies and this has resulted in many data that can be compared directly to human exposures. In turn, the results of these comparisons should provide additional links to studies of lung cancer induced by exposures to other radionuclides inhaled by laboratory animals as discussed below.

One of the primary problems in the use of laboratory animals for the study of radon progeny effects is the comparability of the radiation doses used to those received by humans. This is a particularly important issue for rodents because the radiation doses delivered to the peripheral portions of the lung are much larger than occur for humans, where much greater dose is delivered to the cellular linings of the airways of the lung. Because the types of cells differ between these two regions of the lung, different types of lung tumors are thought to originate from the cells in these regions. In rodents, more adenocarcinomas are seen after exposure to radon progeny than are seen for human exposures, and these tumors are thought to be associated with cells in the peripheral lung. These differences have to be considered in any comparison of the dose-response patterns in rodents to humans for radon progeny (NAS/NRC, 1988).

6.6.5 *Examples of Internally-Deposited Radionuclides for Which Laboratory Animal Data are Available and for Which Links Could be Made to Human Data*

The main value of having studies in laboratory animals that parallel those in human populations is to establish that similar types of effects are seen in the laboratory animals as are seen in humans for the same types of exposure to internally-deposited radionuclides. If this can be established, then it is reasonable to examine how the dose-response relationships are modified for exposures to other radionuclides that differ in their radiation quality or spatial or temporal pattern of dose delivery. These associations make it possible to form a series of links between the few available examples of dose-response relationships in humans to dose-response relationships available in laboratory animals involving exposures to other radionuclides. This Section will discuss possible strategies for using available data to form some of these links. The linkages are addressed on the basis of the organ or tissue where cancer is induced by radiation exposure from an internally-deposited radionuclide. The emphasis is on always maintaining the most direct link possible back to the available human data.

6.6.5.1 *Bone Cancer.* The primary link for bone cancer is to studies of human populations that ingested ^{226,228}Ra occupationally or were injected with ^{224}Ra as a treatment for ankylosing spondylitis or skeletal tuberculosis. Bone cancer has been the primary long-term biological effect seen in these populations. Some life-span studies have been conducted in laboratory animals to strengthen the extension of this dose-response information to other possible human exposure situations involving other radionuclides, radionuclide forms, and routes of exposure. Figure 6.13 illustrates schematically how life-span studies in beagle dogs relate, or are linked to human studies and to each other in terms of bone cancer in beagle dogs.

At the center of the linkages shown are the studies conducted with ^{226}Ra at the University of Utah and the University of California at Davis (Thompson, 1989). These studies, which parallel

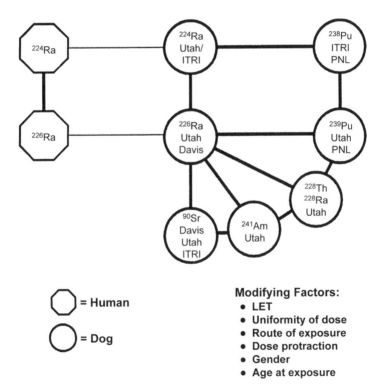

Fig. 6.13. Some important interspecies linkages for bone cancer in humans from ^{224}Ra and ^{226}Ra studies in dogs.
Abbreviations:
Davis = University of California at Davis
ITRI = Lovelace Inhalation Toxicology Research Institute, Albuquerque, New Mexico
PNL = Battelle Pacific Northwest National Laboratory, Richland, Washington
Utah = University of Utah.

human exposure to ^{226,228}Ra, make it possible to compare directly, in the dog, the life-span biological effects from other types of exposure. The earliest comparison of this type, which was between ^{226}Ra and ^{239}Pu, was the basis for establishing the Utah program of life-span studies. By knowing the health effects of ^{226,228}Ra in humans and the ratio of ^{239}Pu-induced health effects to ^{226}Ra-induced health effects in the dog, one can estimate the magnitude of ^{239}Pu-induced health effects in people (Lloyd et al., 2001). Such a toxicity ratio method has been quite valuable in the earlier absence of human dose-response data for internal depositions of ^{239}Pu. The emerging database from workers at the Mayak facility in Ozyorsk, Russia, with internal depositions of ^{239}Pu, is now providing valuable human data on the stochastic effects of internally-deposited ^{239}Pu.

The second side-by-side linkages from human ^{226,228}Ra exposures are the studies of injected ^{226}Ra and chronically ingested ^{90}Sr at Davis (Thompson, 1989). This linkage allows comparisons involving different routes of administration (intravenous versus ingestion) and LET (alpha versus beta).

The other primary linkage between human and dog studies of radionuclide-induced bone cancer involves ^{224}Ra. The study in dogs shown in Figure 6.13 was initiated at Utah and completed at the Inhalation Toxicology Research Institute (ITRI) (Muggenburg et al., 1996a; Thompson, 1989). Because of the short radioactive half-life of ^{224}Ra (3.62 d), much of the alpha radiation is emitted while the ^{224}Ra is on bone surfaces, in contrast to ^{226}Ra, which is primarily a bone-volume seeker. Thus, the results of these studies of two different isotopes of Ra provide a comparison of effects from a surface-seeking and a volume-seeking radionuclide.

The studies at ITRI and PNL of inhaled ^{238}PuO$_2$ have involved many bone cancers from skeletal deposition of the ^{238}Pu after fragmentation and dissolution of the oxide form in the lung (Gilbert et al., 1998; Muggenburg et al., 1996a; Park et al., 1997; Thompson, 1989). The linkages shown provide the opportunity to compare ^{238}Pu-induced bone tumors from two laboratories with results from the ^{224}Ra and ^{226}Ra studies and the bone tumors induced by ^{239}Pu given intravenously at Utah or by inhalation of a soluble form, ^{239}Pu(NO$_3$)$_4$ at PNL.

Other linkages to ^{226}Ra and ^{239}Pu in the Utah program include studies of ^{228}Th, ^{228}Ra, ^{241}Am, and ^{90}Sr (Thompson, 1989). Studies of the beta emitter ^{90}Sr, conducted in three laboratories by three routes of administration (ingestion, inhalation and intravenous injection), provide opportunities for interstudy comparisons as well as comparisons with other studies involving alpha-emitting radionuclides. Other linkages are also available beyond those given as examples here.

Several studies of radionuclide-induced bone tumors in rodents have been performed by laboratories in Europe. Descriptions of many of these studies are now available in a report on the *International Radiobiology Archives of Long-Term Animal Studies* (Gerber *et al.*, 1996). The linkages of these studies are illustrated in Figure 6.14 along with the linkages shown in Figure 6.13 for dogs. These studies provide direct linkages to human studies from rodents exposed to ^{226}Ra and ^{224}Ra at several different laboratories. Modifying factors can be studied based upon these direct linkages with the human data. By comparing studies in rodents and dogs, extrapolations to humans of the effects of modifying factors on bone tumor dose-response relationships can be strengthened by verifying the effects in multiple species.

6.6.5.2 *Liver Cancer.* For a detailed discussion of liver cancer produced by internally-deposited radionuclides in both humans and laboratory animals, the reader is referred to NCRP Report No. 135, *Liver Cancer Risk from Internally Deposited Radionuclides* (NCRP, 2001a). There are two human populations in which liver cancer has been produced as a long-term biological effect of an internally-deposited radionuclide, subjects injected with Thorotrast® and the Mayak workers. The first population comprises groups of people given injections of the radiographic contrast medium Thorotrast® that were discussed in Section 6.5.4.3 (Gossner *et al.*, 1986; Machinami *et al.*, 1999; NAS/NRC, 1988; NCRP, 2001a; Taylor *et al.*, 1989). Because of the very long-term follow-up that has occurred in these studies, almost all of the subjects are now dead.

Linkages for expanding the application of liver cancer risks from the Thorotrast® data in humans are shown in Figure 6.15. In contrast to the bone cancer linkages shown in Figure 6.13, no studies have been conducted with dogs that parallel the human experience with Thorotrast®. However, there are a number of life-span studies with dogs that inhaled, or were injected with, other alpha- or beta-emitting radionuclides that resulted in hepatic degeneration and cancer. In this case, the studies in rodents that were injected with Thorotrast® mentioned above can provide useful linkages between human Thorotrast® studies and the dog studies involving liver cancers produced by other alpha- and beta-emitting radionuclides.

170 / 6. EXTRAPOLATION MODELS

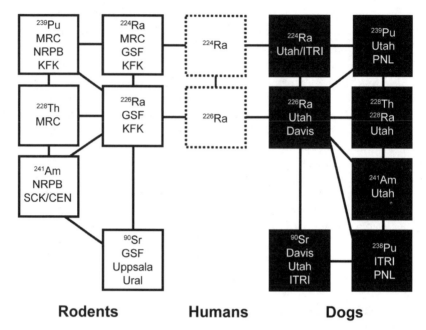

Rodents   **Humans**   **Dogs**

Fig. 6.14. Some possible linkages for studies of bone cancer between human epidemiological and laboratory animal studies from the U.S. and European archives. These linkages provide ways to extend available human data to other potential exposure situations for which few or no human data exist.

Abbreviations:
Davis = University of California at Davis
GSF = Forschungszentrum fur Umwelt und Gesundheit in
   Neuherberg, Germany
ITRI = Lovelace Inhalation Toxicology Research Institute,
   Albuquerque, New Mexico
KFK = Kernforschungszentrum Karlsruhe, Germany
MRC = Medical Research Council in Didcot, United Kingdom
NRPB = National Radiological Protection Board in Didcot, United
   Kingdom
PNL = Battelle Pacific Northwest National Laboratory, Richland,
   Washington
SCK/CEN = Studicenter voor Kernenergie/Centre d'Etude de l'Energie
   Nucleaire in Mol, Belgium
Uppsala = Swedish University of Agricultural Science in Uppsala
Ural = Ural Research Center of Radiation Medicine in Chelyabinsk,
   Russia
Utah = University of Utah

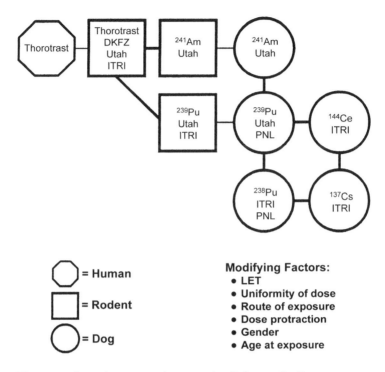

Fig. 6.15. Some important interspecies linkages for liver cancer. Abbreviations:
DKFZ = Deutches Kernforschungszentrum, Heidelberg, Germany
ITRI = Lovelace Inhalation Toxicology Research Institute, Albuquerque, New Mexico
PNL = Battelle Pacific Northwest National Laboratory, Richland, Washington
Utah = University of Utah

The second population, for which a database on ^{239}Pu-induced liver cancer is emerging, comprises workers in the Mayak Production Association in Ozyorsk, Russia (Gilbert et al., 2000; Koshurnikova et al., 1999; Shilnikova et al., 2003). As the Mayak dose-response data become well-established, they can provide a more direct link to studies of plutonium-induced liver cancers in dogs and rodents, and from there to liver cancers induced in dogs by the beta emitters ^{144}Ce and ^{137}Cs and their progeny (Dagle et al., 1996; Hahn et al., 1996; 1999; Muggenburg et al., 1996a; Nikula et al., 1995).

6.6.5.3 *Lung Cancer.* The radiation epidemiology studies in humans in which dose-response relationships for lung cancer have been observed are those of underground miners that inhaled radon and it progeny (NAS/NRC, 1999), Japanese atomic-bomb survivors who were exposed to neutrons and x rays (Pierce et al., 1996), persons exposed to x rays for medical purposes such as Hodgkin's disease (Gilbert et al., 2003), breast cancer (Inskip et al., 1994), and peptic ulcer (Carr et al., 2002), and the emerging database on Mayak workers exposed to external x-ray sources and/or alpha irradiation from inhaled ^{239}Pu (Kreisheimer et al., 2000). The applicability of the studies of x-ray-induced lung tumors in humans to exposure from inhaled radionuclides is questioned because the x-ray exposures were at high-dose rates compared to the much lower-dose rates from radionuclides in the lung.

The applicability of the studies of underground miners to other types of inhaled radioactive particles in the lung has been questioned because exposure to radon and its progeny exposed the epithelium of the airways of the lung to much higher doses than they exposed the deeper parts of the lung that make up the pulmonary region. Exposure to other types of radioactivity deliver more dose to the peripheral pulmonary portions of the lung relative to the dose delivered to the airway epithelium. These two regions of the lung have different types of cells that are thought to give rise to different types of lung tumors. Many of the studies of underground miners are thought to arise mainly from cell types in the epithelium of the airways (Saccomanno et al., 1996).

In another examination of lung-tumor types after radiation exposure, Land et al. (1993a) reported on a comparison of lung tumor types seen in the Japanese atomic-bomb survivors and the uranium miners. They found that the relative frequency of small-cell cancers and adenocarcinomas were very different in these two populations. These differences were accounted for by dose-based relative-risk estimates. They concluded that radiation-induced cancers appeared more likely to be of the small-cell type and less likely to be adenocarcinoma in both populations.

Studies in laboratory animals at COGEMA, PNL, and ITRI provide links between the studies in humans to other types of exposures of the lung from inhaled radionuclides that are important for purposes of health protection. Examples of linkages for understanding how dose-response relationships for lung cancers are modified by various factors are shown in Figure 6.16. These studies in rodents and dogs provide information about how risks of lung cancer may be modified by radiation quality, temporal protraction of dose to the lung, different patterns of spatial delivery of dose to the lung, gender and age at which exposure to radiation started. Many of these studies have been conducted in dogs, but none of

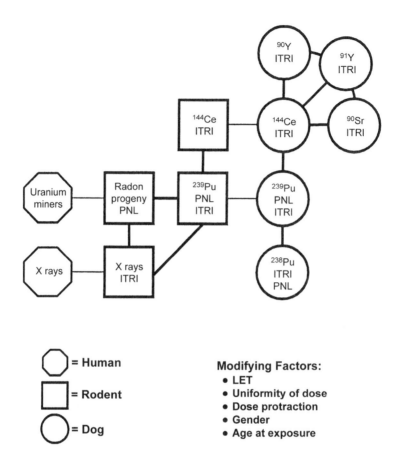

Fig. 6.16. Some important interspecies linkages for lung cancer. Abbreviations:
ITRI = Lovelace Inhalation Toxicology Research Institute, Albuquerque, New Mexico
PNL = Battelle Pacific Northwest National Laboratory, Richland, Washington

these are of sufficient size to provide a direct linkage to human exposures. Linkages have been provided by studies in rats exposed to radon and its progeny at PNL and COGEMA (Cross and Monchaux, 1999; Heidenreich *et al.*, 2004; Kaiser *et al.*, 2004; NAS/NRC, 1988) and to x rays at ITRI (Lundgren *et al.*, 1992a; 1992b) for a direct comparison to the human studies. These studies are then linked to dog studies by additional studies of rats that inhaled relatively insoluble particles containing ^{239}Pu (Lundgren *et al.*, 1995; Sanders and Lundgren, 1995; Sanders *et al.*, 1993a;

1993b) or ^{144}Ce (Lundgren *et al.*, 1992c; 1996). These radioactive particles are retained primarily in the peripheral parts of the lung where they irradiate the lung over a substantial portion of the life span of the rat. These studies provide an understanding of how the differences in dose delivery between radon progeny and radionuclides deposited in the peripheral pulmonary region affect the risks of lung cancer from radiation.

To understand relationships between rat and dog more fully for developing lung cancer from radiation exposure, studies are available in which both species inhaled insoluble radioactive particles containing ^{239}Pu or ^{144}Ce (Boecker *et al.*, 1994; Hahn *et al.*, 1999; Park and Staff, 1993; Thompson, 1989). In these studies, the dose was received chronically, primarily to the peripheral regions of the lung where the particles were retained for prolonged periods.

Results of these studies then provide linkages from ^{239}PuO$_2$ through the studies with ^{144}Ce in fused aluminosilicate particles to studies of other beta-emitting radionuclides also inhaled in an insoluble fused aluminosilicate particle matrix; ^{90}Y, ^{91}Y, and ^{90}Sr. By using these beta emitters with different radioactive half-lives, it was possible to vary the duration of time during which the lung was chronically irradiated. This resulted in a temporal protraction of the period when the lung was irradiated from a period of about one week for ^{90}Y, about three months for ^{91}Y, about eight months for ^{144}Ce, and several years for ^{90}Sr. These studies showed that delivering the dose from a beta emitter more quickly increased the effectiveness in producing lung tumors (Boecker *et al.*, 1997).

The studies of inhaled ^{239}PuO$_2$ in dogs provide a basis for evaluation of the spatial distribution of dose in the pulmonary region of the lung. In the studies at ITRI, three different sizes of monodisperse particles (all of the inhaled particles are approximately the same size) were used (Boecker *et al.*, 1994). These particles differed in size by factors of about two between each other and thus by a factor of about 50 in activity per particle from the smallest to the largest. This had the effect of providing much higher radiation doses around the larger particles. The distribution of dose across the lung was such that only ~10 % of the lung tissue was irradiated by the larger particles at any time, whereas all of the lung tissue was within range of the smaller particles.

The studies at PNL with ^{239}Pu dioxide complemented these studies by using polydisperse particles (particles with a wide range of sizes), which is the distribution of particle shapes likely to be found in an actual accident (Park and Staff, 1993).

The studies with inhaled ^{238}PuO$_2$ at PNL and ITRI were intended originally to extend the range of particle activities to those that would produce an even larger dose around each particle, because ^{238}Pu is an alpha emitter with a shorter half-life than ^{239}Pu (~90 y compared with ~24,000 y). However, the activity per particle was high enough to cause radiation damage to the particles themselves. This caused the particles to fragment and dissolve from the lung by ~300 d after exposure in the ITRI study and somewhat later in the PNL study. Despite this unexpected behavior, the particles were in the lung long enough to deliver sufficient radiation dose to the lung to produce a high incidence of lung tumors in both studies. The ^{238}Pu dissolved from the lung was translocated to the liver and skeleton and led to a high incidence of tumors in these organs in contrast to the studies with ^{239}Pu that did not dissolve from the lung. Thus, the ^{238}Pu studies in dogs provide an additional temporal pattern of dose to the lung for alpha-emitting radionuclides (Muggenburg *et al.*, 1969a; Park *et al.*, 1997).

6.6.6 *Examples of Linking Risks from Laboratory Animals to Human Data*

6.6.6.1 *Bone Cancer.* Many of the earliest comparisons that linked results of studies in laboratory animals to human data used a toxicity ratio method as discussed by Mays *et al.* (1969). The general thrust of the approach is that one might expect the ratio of toxicities for ^{239}Pu to ^{226}Ra in humans to equal the ratio of toxicities of ^{239}Pu to ^{226}Ra in a laboratory animal species. This was an important concept in the experimental design of the studies of bone-seeking radionuclides in beagle dogs conducted at the University of Utah (Mays *et al.*, 1986a). By determining the toxicity ratio of skeletally deposited ^{239}Pu to ^{226}Ra in dogs and knowing the toxicity of skeletally deposited ^{226}Ra in humans, one could estimate the toxicity of skeletally deposited ^{239}Pu in humans according to this relationship:

$$^{239}\text{Pu toxicity}_{man} = \frac{^{236}\text{Ra toxicity}_{man} \times {^{239}}\text{Pu toxicity}_{dog}}{^{226}\text{Ra toxicity}_{dog}}. \qquad (6.20)$$

The validity of the toxicity ratio method for extrapolation from animals to humans hinges on the assumption that this ratio is approximately constant for the same radionuclide in two different species. Evans *et al.* (1969) emphasized this fact and pointed out that this assumption could be checked out with ^{226}Ra and ^{228}Ra in dogs and humans.

In early usage, no specific definition was given for how to quantify "toxicity." Early practitioners used, for example, "average time to death with osteosarcomas." More recent analyses have used slopes of fitted linear dose-response relationships.

Figure 6.17 illustrates results of the toxicity ratio in beagle dogs injected with ^{226}Ra or ^{239}Pu (Lloyd et al., 1993). Linear fits were made to grouped data on the incidence of bone cancer for different average doses to the skeleton. From the two lines shown, it is clear that the fitted line for ^{239}Pu is much steeper than that for ^{226}Ra, reflecting an increased carcinogenicity of ^{239}Pu relative to ^{226}Ra by a factor of 16 when the average radiation dose to the skeleton is used (Figure 6.17).

Bone cancer ratio values [±1 SD (standard deviation)] for other radionuclides relative to ^{226}Ra in dogs computed by Lloyd et al. (1994) are:

- ^{224}Ra (50 injections), 16 ± 5
- ^{228}Th, 8.5 ± 2.3

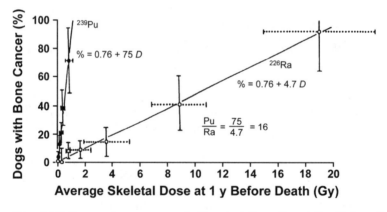

Fig. 6.17. Comparisons among studies; toxicity ratios at the University of Utah (Lloyd et al., 1993). Percent of dogs in each dose level with bone cancer as a function of average skeletal dose at 1 y before death for beagles given a single injection of ^{239}Pu or ^{226}Ra citrate at ~1.5 y of age. Vertical error bars show plus or minus the estimated uncertainty in the percent of dogs with tumors (the error bar that extends to >100 is not shown in its entirety; error bars do not extend beyond the area of the symbol are not shown). Horizontal error bars show ±1 SD in skeletal dose for the various dose levels (error bars that do not extend beyond the area of the symbol are not shown). The equation for ^{239}Pu represents the dose-response relationship below ~1.3 Gy, and the equation for ^{226}Ra represents the dose-response relationship below ~20 Gy. Standard deviation for the two risk coefficients (not shown in the figure) were derived from the curve-fitting procedure (NCRP, 1991) and were ±22.5 for ^{239}Pu, and ±0.47 for ^{226}Ra. The SD in the ratio in the two risk coefficients (±5) was calculated as the relative root mean square of the SDs of the point estimates.

- ^{241}Am, 6 ± 0.8
- ^{228}Ra, 2 ± 0.5
- ^{90}Sr (>40 Gy), 1 ± 0.5
- ^{90}Sr (5 to 40 Gy), 0.05 ± 0.03
- ^{90}Sr (<5 Gy), 0.01 ± 0.01

The value for ^{224}Ra given by repeated injections is the same as that for ^{239}Pu, a result consistent with the mixed bone-surface/bone-volume deposition characteristics of this short-lived radionuclide. Decreasing ratio values are seen for the other radionuclides listed. For ^{90}Sr, the dose-response curve was nonlinear. Under these conditions, values could be computed at different doses, and three are given for illustrative purposes. This toxicity ratio method is easy to compute and understand, and provides one way of comparing the carcinogenicity of different radionuclides in the same species.

The suggestion by Evans *et al*. (1969) to consider collateral relevant information was put to use about two decades later when the risk from plutonium isotopes was estimated with a hierarchical model using information for several species and radionuclides (DuMouchel and Groer, 1989). This hierarchical modeling approach used Bayesian techniques for parameter estimation. This statistical approach added the important feature of a quantitative description of uncertainty to this extrapolation scheme based on relative toxicity. The meta-analysis published by DuMouchel and Groer unified data from humans, dogs and rats to derive an estimate of the bone cancer risks in humans for exposure to ^{239}Pu. An analysis of this type was also used in the NAS/NRC (1988a) report. This innovative analysis used Bayesian statistical method to combine summary information on the dose groups in 15 studies to estimate a linear dose-response coefficient and its variance for each study. The variation in slope among the studies was compared to the within-study sampling variation. This analysis suggested that extrapolations across species could be made only with an accuracy of a factor of two to four even when within-study sampling variation is small. A risk of bone cancer induced by ^{239}Pu is estimated to be 300 cancers per 10^4 person-Gy with a 95 confidence interval of 80 to 1,100.

Raabe (1989; 1994) has devised an alternative method for studying dose-response relationships and has applied it to bone cancer (Figure 6.18) and lung tumors. The dose-response relationships in this method describe changes in survival time as a function of the average dose rate, rather than relating the tumor incidence or prevalence to the radiation dose as has been done in most other

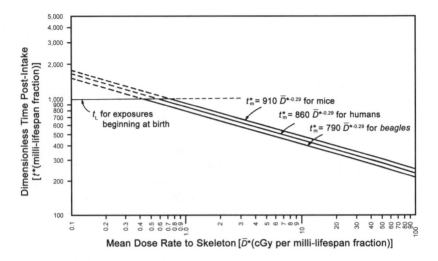

Fig. 6.18. Distribution of deaths for beagle dogs and predicted distribution of deaths for people exposed to ^{226}Ra along with human data from the U.S. radium cases. Line of median dimensionless time (t_m^*) of fitted median risk ($z = 0$) for each species (Raabe, 1989) and t_L the estimated median times of spontaneous deaths among unexposed controls (Raabe et al., 1983).

approaches. While this method has some intuitive appeal, it relies mainly on extrapolation from very high doses to a lower-dose range. At the lower doses in the studies and experiments analyzed by Raabe, there is little detailed evaluation of the fit of the model or consideration of alternative functions for the dose response in the low-dose region. This is the part of the dose range in these studies that is of greatest interest because the dose range is closest to doses at which people are likely to be exposed. Small differences in survival or tumor incidence in this lower-dose range may have large consequences when extrapolated to humans. Because the methods cited by Raabe have not been used by other authors (Raabe, 1994), it is difficult to evaluate whether they are reliable predictors for low-dose effects from internally-deposited radionuclides. The reliability and applicability of these methods need to be established to understand how they might be used to extrapolate to human health effects.

A fourth approach that has been used to analyze data from the life-span studies involves PHM. Figure 6.19 illustrates this approach using bone cancer data in beagle dogs that inhaled relatively soluble forms of ^{90}Sr or ^{238}Pu at ITRI or were injected with

6.6 EXTRAPOLATION OF RESULTS / 179

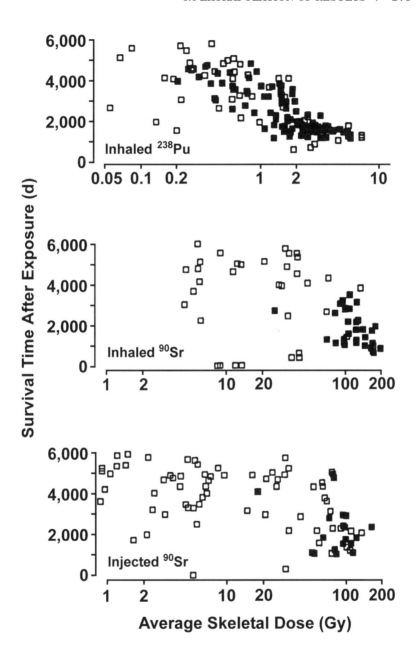

Fig. 6.19a. Proportional hazards modeling using bone cancer data in beagles that were injected with ^{90}Sr citrate or that inhaled ^{90}SrCl$_2$ or ^{238}PuO$_2$ (filled symbols are bone tumors) (Griffith et al., 1995).

180 / 6. EXTRAPOLATION MODELS

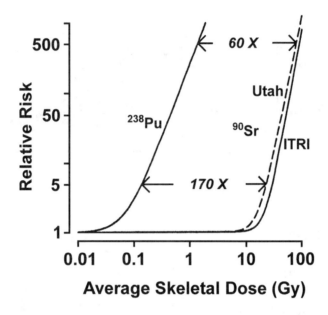

Fig. 6.19b. Estimates of the relative risk in a proportional hazards model for bone cancer in dogs that inhaled ^{238}PuO$_2$ at ITRI, dogs that inhaled ^{90}SrCl$_2$ at ITRI, and in dogs injected with ^{90}Sr citrate at Utah. The increase in average skeletal dose for a relative risk of 5 and 500 between inhaled ^{238}PuO$_2$ and the two soluble forms of ^{90}Sr that were inhaled or injected is shown.

^{90}Sr at the University of Utah (Griffith *et al.*, 1995). The skeleton was a major site of radionuclide deposition in all three studies. Once deposited in the skeleton, it was retained very tenaciously and the cumulative dose continued to increase every year until death occurred. The skeletal dose to each dog was the total received from the time of radionuclide exposure to the first radiographic diagnosis of a bone tumor in that dog. The plots shown in Figure 6.19a illustrate the relationship between dose and time to bone cancer in these three studies. This method considers the time-dependent nature of the dose received from an internally-deposited radionuclide. Using the Cox proportional hazards model (Cox and Oakes, 1984), these results can then be computed as relative risks based on the hazard rate (age-specific rate) for bone tumors in control dogs, resulting in the curves shown in Figure 6.19b. From these three curves, which were based on average skeletal doses, the alpha emissions from ^{238}Pu were over 60 to 170 times more effective than the beta emissions from ^{90}Sr-^{90}Y. If an endosteal surface dose were used to describe the alpha radiation from the surface-deposited ^{238}Pu, the relative effectiveness would be divided by about a factor of 10. This is based on the ICRP bone dosimetry model described in ICRP Publication 30 (ICRP, 1979). Surface deposition would lead to a higher calculated dose and therefore a reduced effectiveness per unit of dose. Both of these results can be compared with the current ICRP radiation weighting factors for $\alpha/\beta = 20/1 = 20$.

Also, this example illustrates the similarity of dose response for studies involving two different routes of exposure, inhalation and injection of ^{90}Sr, performed at two different laboratories. The dose-response curves suggest an effective threshold for bone tumors because both involve a power of the dose, 4.2 for ^{90}Sr and two for ^{238}Pu. When a linear slope was added to the curves for either ^{90}Sr or ^{238}Pu, the linear parts were not statistically significant for either radionuclide.

This analysis has also been extended to the study of injected ^{224}Ra. The results of this analysis are shown in Figure 6.20 for the dogs receiving 50, 10 and 1 weekly injections. ^{224}Ra has a 3.6 d half-life so these injection schedules involve protraction of the dose to bone of ~1 y, 3 months, and 10 d. This analysis illustrates the reverse dose-rate effect for bone from protracting the radiation dose, in a manner similar to that for humans injected with ^{224}Ra as a medical treatment. Also the group receiving 50 weekly doses of ^{224}Ra has a similar dose-response curve to that for ^{238}Pu, suggesting that protraction of the dose beyond 1 y may not increase its effectiveness.

6.6.6.2 *Lung Cancer.* A PHM analysis comparing lung tumors induced by radon progeny and ^{239}PuO$_2$ in rats was published by Gilbert *et al.* (1990). They found major differences in the dose response for these two alpha emitters in inducing lung tumors when the tumors were assumed to be rapidly fatal, but the responses were similar when the tumors were assumed to be incidental to the death of the rats. This same group also published a more extensive analysis of the lung tumors in rats exposed to radon using the same types of methods (Gilbert *et al.*, 1996). Overall they found comparable risks to those estimated for humans, 237 lung cancers per 10^6 rats per WLM. The data below 1,000 WLM were consistent with a linear dose response. Evidence for an inverse dose-rate effect was limited primarily to exposures exceeding 1,000 WLM.

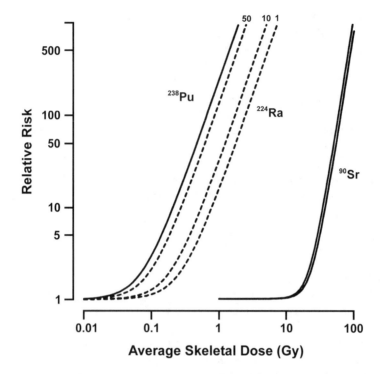

Fig. 6.20. Proportional hazard modeling estimates of relative risk for induction of bone tumors in beagles from (1) a single inhalation exposure to ^{238}PuO$_2$ at ITRI; (2) 50, 10 or 1 weekly injections of ^{224}RaCl$_2$ at Utah; (3) a single intravenous injection of ^{90}Sr citrate at Utah; and (4) a single inhalation exposure to ^{90}SrCl$_2$ at ITRI.

7. Summary

7.1 Introduction

This NCRP Report deals with the procedures by which radiation-induced, long-term or chronic injury (*e.g.*, a reduction of life expectancy or increases in age-specific cancer mortality) measured in laboratory animal species, can be used to estimate risks to humans. The data for such extrapolation procedures are discussed in detail. The issue of extrapolation is complicated because data are incomplete, and there is some uncertainty about the applicability of existing models since they are not tissue-specific. Nonetheless, it has been possible to develop what appear to be useful and relevant extrapolations for certain situations. This Report focuses on those situations.

Current guidelines on limits of exposure to ionizing radiation issued by ICRP in Publication 60 (ICRP, 1991) and by NCRP in Report No. 116 (NCRP, 1993) rely primarily upon the use of epidemiological data to estimate the risk of cancer in exposed human populations. The majority of the data used for this purpose has been obtained from the Japanese survivors of the atomic bombs, with additional data provided by studies on occupationally and medically exposed individuals. However, the ability of epidemiology studies to determine cancer risk in humans at low radiation exposure levels, below ~50 mSv, is limited. In addition, there are only limited data available on the effects in humans of protracted radiation at low exposure dose rates, and of exposure to high-LET radiation (*e.g.*, from neutrons and heavy charged-particle radiation). This lack of human data has severely limited the reliability of estimates of DREF and the radiation weighting factor used for evaluating cancer risks in human populations exposed to ionizing radiation.

This Report undertakes an extensive review of data on radiation-induced cancer in specific major organ systems in several species of laboratory animals. Extrapolation models are described that can be used under certain conditions to apply the extensive data base from laboratory animal studies to the evaluation of human radiation risks. Data from laboratory animals are also used for the estimation of genetic risk factors for radiation-induced cancer and for estimating values of DREF and radiation weighting factor.

7.2 Summary

The following is a brief summary of the primary topics of discussion in this Report on extrapolation of radiation risks from nonhuman species to humans. After the Executive Summary and Introduction to this NCRP Report, four main topics are sequentially addressed in detail. Interested readers can refer to these specific sections of the Report for elaboration, including related specific recommendations and conclusions. These topics are:

7.2.1 *History of Extrapolation from Nonhuman Experimental Systems to Humans*

The histories of genetic and carcinogenic studies and actuarial life-table models are traced from the early 19th century work of Gompertz through the progressive understanding that was acquired during the 20th century of several important aspects of cancer risk and life shortening, including: (1) the relationship between genetic mutations and cancer risk, (2) various species-dependent factors that influence cancer risk and life span, and (3) the application of actuarial life tables to describe radiation-induced life shortening among commonly used laboratory animals and humans.

The important point made in this Report is that a comprehensive understanding of cellular and molecular genetic effects of radiation as a basis for carcinogenesis must be combined with actuarial models of aging in several species of mammals to allow a meaningful interspecies comparison to be made of radiation-induced life shortening. Without this underlying basis, predictions of radiation-induced carcinogenesis and life shortening in humans cannot be made with confidence.

Every example of an interspecies prediction of radiation-induced mortality contained within this Report was, however, made in the absence of an understanding of cellular and molecular genetic effects of radiation. These latter types of effects are important for providing a degree of confidence to the use of the so-called biological models, but they were not critical, as stated above, for the empirical models. Clearly, continued research in these areas is needed to further the bases underlying predictive capability.

Within the relative risk framework, the dose or dose-rate-related risks are expressed as multiples of the mortality risks observed in the control population. The control laboratory animal population has no radiation risks, but does express the age-specific mortality risks of growing old. This reality for the controls suggests that it is necessary to understand the mortality risks of aging in order to understand the context of the enhancement/acceleration of these risks caused by radiation.

7.2.2 *Cells of Origin of Cancer in Different Animal Species*

The current understanding of the origins of carcinogenesis based on studies in laboratory animals is discussed for seven tissue and organ systems (hematopoietic system, lung, breast, thyroid, skin, GI tract, and bone). These systems were chosen because of their frequent occurrence in laboratory animals and humans following exposure to external sources of radiation, and from internally-deposited radionuclides. For cancers in each of these target tissues and organs, a summary was given of the cells of origin and other physiological factors that influence radiation-induced cancer risk in various mammalian species, primarily rodents, dogs and humans. Based on these considerations, conclusions are drawn on the feasibility of extrapolating radiation cancer risk measured for these seven tissues and organs in laboratory animals to humans.

7.2.3 *Radiation Effects at the Molecular and Cellular Levels*

The mechanisms of radiation induction of chromosome aberrations and gene mutations are described for humans and laboratory animals. A description is also provided of radiation effects at the molecular and cellular levels (primarily DNA damage and repair mechanisms, cytogenetic effects, and modifications such as oncogene activation and the inactivation of suppressor genes), and a comparison of these effects across various species including humans. It is concluded that radiation-induced mutation rates across species, when differences in cellular DNA content are accounted for, differ at most by a factor of two. This fact, coupled with extensive data indicating that the underlying mechanisms of tumor induction in rodents and humans are similar, lends support to the credibility of extrapolating cancer risks measured in laboratory studies with rodents to humans. This conclusion is tempered, however, by the recognition that the specific gene alterations related to a particular type of cancer often differ across species. In addition, there are species-specific host factors that can alter the probability of tumor development from irradiated cells.

7.2.4 *Extrapolation Models*

An analysis is provided of the reliability of extrapolation models for relating cancer incidence and life shortening measured in laboratory animals to the same endpoints in humans. The correlation of chemical toxicity observed for many mammalian species is described, as an example, where extrapolation models can be successful if they are based on common physiological traits of the different species. Examples are then provided of several predictions of cancer and life shortening in humans based on data from laboratory animals; for example, the prediction of survival curves for Japanese survivors of atomic-bomb radiation from the results of studies on mice exposed to single doses of gamma radiation.

This Report also provided a successful extrapolation of radiation-induced mortality from the mouse to the dog for continuous exposures. The successful predictions from mice to dogs for continuous exposures, and mice to humans for single exposures, provide a basis for optimism about the extrapolation from mice to humans for continuous exposure.

A Gompertz based approach, which estimates days of life lost per unit dose may be an old technique, but it should not be ignored, because it retains utility. To quote from Carnes *et al.* (2003) the paper upon which the NCRP example was derived, "An examination of a historical metric of radiation-induced injury from an interspecies perspective demonstrated that readily available information permits defensible predictions of radiation-induced mortality to be made for species lacking reliable exposure information. All that is needed is a dose-response slope for any species with reliable exposure data, and an estimate of the Gompertz rate parameter of a representative control population for the species of interest."

An important point of this Report is that there was no adjustment for host factor differences in radiation sensitivity. In other words, the use of such an approach suggests that, at doses below those causing acute effects, host factor differences were not important, at least for the larger combined pathology endpoints like life shortening, intrinsic mortality and solid-tissue tumor mortality.

Other examples where extrapolation of cancer risks appear to be reasonable include: (1) the estimation of a DREF from mammary cancer mortality in female mice exposed to gamma radiation at low-dose rates, and (2) the risk of bone cancer in dogs exposed to internally-deposited radionuclides such as radium and strontium. Limitations on the reliability of such extrapolations imposed by species-specific risk factors and age at exposure are also discussed.

This Report includes, where appropriate, a set of recommendations and conclusions specific and relevant to each section. It was deemed important to have these conclusions as part of the relevant section because of the proximity of the supporting text. Interested readers should consult the particular section for details supporting the conclusions and recommendations. For example, the feasibility of interspecies comparisons of cancer risk and life shortening, followed by a set of recommendations for future research are presented in Section 1. These recommendations include: (1) the need to obtain and utilize molecular mechanisms underlying the cellular events that lead to cancer in various animal species, (2) the need to archive data in a manner that is amenable to meta-analysis, and (3) the need to acquire more information on the carcinogenic effects of heavy charged-particle radiation such as that encountered by astronauts during deep-space missions.

Environmental health, in general, is necessarily dependent upon animal toxicology and its application in quantitative risk estimation. The study of ionizing radiation effects has the luxury of both varied animal studies and extensive human data with good exposure estimates. There is reasonable agreement between animals and humans as long as biological considerations are not ignored. The key issue is the degree of uncertainty in animal extrapolation. This latter aspect will be improved with more extensive analysis of existing data coupled with development through improving technology of our understanding of the biological issues of radiation and disease.

Glossary

adenocarcinoma: A malignant neoplasm of epithelial cells in glandular or gland-like pattern.

adenomatous polyposis coli: Adenomas with colon polyps.

apocrine: Type of secretion in which the apical portion of the secretory cells is shed and incorporated into the secretion.

apoptosis: Deletion of cells by fragmentation into membrane-bound particles that are phagocytosed by other cells; often referred to as "programmed cell death."

ataxia-telangioectasia (A-T): A slowly progressive multisystem disorder appearing at the onset of walking, including but not limited to dilation of small or terminal vessels of a body part, and recurrent infections.

autocrine: Indicating self-stimulation through production of a factor and a specific receptor for it.

basal cell nevus syndrome: Pertaining to basal cell nevi with development of basal cells carcinomas in adult life (see **nevi**).

base pairs (of DNA): Normally, a base pair involves a pyrimidine (cytosine, thymine) cross-linked to a purine (adenine, guanine), with each component (the purine, the pyrimidine) part of two antiparallel strands winding "right handed" in a double-stranded helix of DNA.

Bayesian: A concept named after an 18th century Presbyterian minister (Thomas Bayes); the concept allows one to start with what one already knows or believes and then to determine how new information changes one's confidence in that belief (Malakoff, 1999; Sontag *et al.*, 1998; Prelec, 2004).

biodosimetry: Measurement of biological response as a surrogate for radiation dose.

bystander effects: A term related to the fact that gene products of transfected DNA can pass from transfected cells to neighboring cells through cell-gap junctions. In radiobiology, the term is used to describe the effects on cells in which the energy had not been directly deposited. In most instances. the cells so affected were neighbors of the cells directly impacted by the radiation.

carcinoid tumor: A usually small, slow-growing neoplasm composed of islands of cells, usually occurring in the gastrointestinal tract.

Clara cell: A rounded, club-shaped, nonciliated cell protruding between ciliated cells in the bronchiolar epithelium. Named after Max Clara, an Austrian anatomist, 1899.

comedo: A dilated hair follicle infundibulum filled with keratin, squamae, bacteria and sebum.

cribriform: Sieve-like, with many perforations.

cytokines: Hormone-like low molecular weight proteins that regulate the intensity and duration of immune responses, and play a role in cell-to-cell communication.

***Drosophila*:** Genus name for the common fruit fly.

endonuclease: An intracellular enzyme that cleaves polynucleotides (nucleic acids) at interior bonds, thus producing poly- or oligonucleotide fragments of varying size.

enolase: An enzyme catalyzing the reversible dehydration of 2-phospho-D-glycerate to phosphoenolpyruvate and water: a step in both glycolysis and gluconeogenesis.

Ewing's sarcoma: A malignant neoplasm, occurring usually before age 20, about twice as frequently in males relative to females, and, in ~75% of the patients, involves bones of extremities. Named after James Ewing, a U.S. pathologist, 1866 to 1943.

exon: A portion of DNA that codes for a section of the mature messenger RNA from that DNA, and is therefore expressed ("translated") into protein at the ribosome.

extrinsic mortality: Causes of death that originate from the environment outside the individual (*e.g.*, accidents, homicide, suicide, infectious and parasitic diseases, famine); its counterpart is intrinsic mortality.

fibrosarcoma: A malignant neoplasm derived from deep fibrous tissue characterized by bundles of immature proliferating fibroblasts that invade locally and metastasize *via* the blood stream.

gene nomenclature: The gene nomenclature in this Report attempts to conform to the new (emerging) international standards. Each major organism (*e.g.*, humans, rats, mice, etc.) has an international committee overseeing nomenclatural aspects of gene identification. Thus, each major organism has its own genetic nomenclature. Each of the nomenclatures (*e.g.*, "gene nomenclature humans," or "gene nomenclature mice") can be readily accessed through normal web search engines. In general, human genes are all italicized capital letters (*e.g.*, *MYC*) while rats and mice genes have only the first letter normally capitalized and all are italicized (*e.g.*, *Myc*). Each gene nomenclature roster contains the earlier-used gene symbols(s) in addition to its present nomenclatural designation, and in some instances, there can only be referral to an earlier (synonym) symbol.

Gompertz function/distribution: A statistical distribution developed in 1825 by the British actuary Benjamin Gompertz to describe age-specific mortality risks. Gompertzian mortality is characterized by a linear increase in age-specific death rates on a plot where the horizontal axis (age) is arithmetic and the vertical axis (death rate) is semilogarithmic.

Gorlin's syndrome: Synonymous with "basal cell nevus syndrome."

gray (Gy): The SI unit of absorbed dose of ionizing radiation, equivalent to 1 Joule kg^{-1} of tissue; 1 Gy = 100 rads

half-life: A period of time in which the radioactivity or number of atoms of a radioactive substance decreases by half.

hazard function, and hazard rate: The derived age-specific probability of death or occurrence of an adverse event; also known as the instantaneous failure rate or force of mortality.

heavy ions: Synonymous with heavy charged particles, heavy nuclei, high-Z particles, or HZE-particles.

Hodgkin's disease: A type of malignancy of lymphoid cells, named after a British physician, Thomas Hodgkin, 1798 to 1869.

interspecies: Between or among different species (*e.g.*, dogs, cats, humans).

intraspecies: Within the same species (*e.g.*, different strains of rats).

intrinsic mortality: A collection of causes of death thought to arise from processes within the body (*e.g.*, genetic diseases, both germ line and somatic; spontaneous and heart diseases, as well as the degenerative diseases of aging). Its counterpart is extrinsic mortality.

Langerhans' cells: Dendritic, clear cells in the epidermis, containing distinctive granules that appear rod-like or racket-shaped but lack tonofilaments, melanosomes and desmosomes. Named after a German anatomist, Paul Langerhans, 1847 to 1888. A *tonofilament* is a structural cytoplasmic protein that is particularly well-developed in the epidermis. A *melanosome* is a generally oval pigmented granule produced by melanocytes. A *desmosome* is a site of adhesion between two epithelial cells.

Lieberkuhn, crypts of: Intestinal glands, named after a German anatomist Johann N. Lieberkuhn, 1711 to 1756.

Li-Fraumeni syndrome: Familial breast cancer in young women, with soft-tissue sarcomas in children and other cancers in close relatives.

ligation: Binding.

melanoma: A malignancy derived from cells capable of forming melanin, most commonly in the skin.

menarche: The onset of menstrual function (menses).

menopause: The cessation of menstrual function (menses).

meta-analysis: The process of using statistical method to combine the results of different studies; "meta" is a prefix denoting "after."

metastasis: The shifting of a disease or its local manifestations from one part of the body to another. In cancer, the spread of the disease from a primary site results from dissemination of tumor cells.

myelodysplasia process: An abnormality in the development of the spinal cord, especially the lower part.

nevi (nevus): A circumscribed malformation of the skin.

nulliparous: Never having borne children.

oncogene: Genes that under normal circumstances code for proteins associated with normal cell growth, but may foster malignant processes if mutated or activated.

pancytopenia: A pronounced reduction in circulating erythrocytes, white blood cells, and platelets.

pluripotent: Having ability to affect more than one organ at a time.

protooncogenes: A gene conserved long on the evolutionary scale present in the normal human genome, that appears to have a role in normal cellular physiology and is often involved in regulation of normal cell growth or proliferation; however, as a result of somatic mutations these genes may become oncogenic.

radiation weighting factor: A factor used for radiation-protection purposes that accounts for differences in biological effectiveness among different radiations. The radiation weighting factor is independent of the tissue weighting factor.

sebaceous: Oily or oil-containing.

strain-dependent differences: Usually refers to differences in responses between or among different genetic lines (*i.e.*, strains) of the same species, but it sometimes includes differences between different species; context determines which usage is intended. Sometimes referred to as host factor differences.

T-cell lymphomas: An acute or subacute disease associated with the human T-cell virus.

tissue weighting factor: A factor used for radiation-protection purposes that accounts for differences in biological effectiveness between different radiations. The radiation weighting factor is independent of the tissue weighting factor.

toxicity ratio method: relative effectiveness per gray of average skeletal dose.

upregulate: Opposite of down-regulate.

von Hippel-Lindau syndrome: A type of phacomatosis, consisting of hemangiomas of the retina; associated with hemangiomas of the cerebellum and walls of the fourth ventricle; sometimes associated with cysts or hamartomas of the kidney, adrenal or other organs.

Wilm's tumor syndrome: A malignant renal tumor of young children.

xeroderma pigmentosum: An eruption of exposed skin occurring in childhood and characterized by photosensitivity with severe sunburn in infancy and subsequent emergence of dermal anomalies including atrophic lesions and solar keratoses that undergo malignant change at an early age.

Symbols and Acronyms

ALL	acute lymphocytic leukemia
AML	acute myelogenous leukemia
ANLL	acute nonlymphocytic leukemia
AUC	area under the curve (in a plot of data, a type of data integration)
BCC	basal-cell carcinoma
bd	base damage, as in damage to the bases of DNA; DNA has four bases, two of which are purines (purine, adenine) and two of which are pyrimidines (cytosine, thymine)
CML	chronic myelogenous leukemia
del	deletion
DMBA	dimethylbenzanthracene
DNA	deoxyribonucleic acid; the type of nucleic acid containing deoxyribose as the sugar component and found principally in the nuclei (chromatin, chromosomes) and mitochondria of animal and plant cells, usually loosely bound to protein
D_0	a term relevant to the "target theory" of radiobiology, and generally relevant to sparsely ionizing radiations such as x or gamma rays. D_0 is defined as 1/e (37 %) survival in the linear portion of a dose-response survival curve. D_0 is expressed in units of dose. In theory, an D_0 there will be an average of one lethal "hit" by the ionizing particle per vital target, but because the radiation is "sparsely ionizing" some targets will not be "hit," other targets will be "hit" once, and some targets "hit" more than once.
DDREF	dose and dose-rate effectiveness factor, based upon the fact that different doses and dose rates of ionizing radiation can have different degrees of bioeffectiveness
DREF	dose-rate effectiveness factor, based upon the fact that different dose rates of ionizing radiation can have different degrees of bioeffectiveness
dsb	double-strand breaks in DNA
GI	gastrointestinal
GM-CFU	granulocyte, monocyte – colony forming unit
HOX or *Hox*	see Glossary, "gene nomenclature," for explanation
LD_{10}	a dose of ionizing radiation lethal to 10 % of the irradiated population of organisms
LET	linear-energy transfer; the energy lost per unit length of track of a primary ionizing particle

MAS	mean after-survival (*i.e.*, the mean surviving fraction of cells or individuals after an exposure at a potentially life-threatening level)
MML	mouse myelogenous leukemia
MST	median survival time
MTD	maximum tolerated dose
MYC or *Myc*	see Glossary, "gene nomenclature," for explanation
Nbs1	Nijmegen breakage syndrome gene; a genetic locus postulated to be associated with the repair of DNA double-strand breaks in higher vertebrate cells
NHEJ	nonhomologous end-joining of DNA double-strand breaks
PHM	proportional hazards modeling
preCFUs	precolony forming unit, spleen
PTC1, PTC3	each a distinct rearrangement of the *ret* oncogene
RBE	relative biological effectiveness of a particular treatment to some other treatment
ret	an oncogene
RFM	a strain of mouse
SCC	squamous-cell carcinoma or small-cell cancer
SCLC	small-cell lung carcinoma
ssb	single strand breaks in DNA
TATA	a human binding protein, which regulates certain transcription processes
TD_{25}	a tumorigenic dose producing a targeted effect (*e.g.*, cancer) in 25 % of the subjects who otherwise would not be affected by the cancer
TEB	terminal end buds
TGB	thyroglobuline
TSH	thyroid stimulating hormone
V(D)J	a specific double DNA break and subsequent recombination that occurs in developing B- and T-lymphocytes to provide the basis for the antigen binding diversity of the immunoglobulin and T-cell receptor proteins
WL	working level that is any combination of radon daughters in 1 L of air that results in 1.3×10^5 MeV of alpha energy
WLM	working level month: exposure from inhaling air with a concentration of 1 WL of radon daughters for 170 h

References

AALEN, O. (1976). "Nonparametric inference in connection with multiple decrement models," Scand. J. Statist. **3**, 15–27.

AARONSON, S.A. and TRONICK, S.R. (1985). "Transforming genes of human malignancies," pages 35 to 49 in *The Role of Chemicals and Radiation in the Etiology of Cancer, Carcinogenesis*, Huberman, E. and Barr, S.H., Eds. (Raven Press, New York).

ABOUSSEKHRA, A., BIGGERSTAFF, M., SHIVJI, M.K., VILPO, J.A., MONCOLLIN, V., PODUST, V.N., PROTIC, M., HUBSCHER, U., EGLY, J.M. and WOOD R.D. (1995). "Mammalian DNA nucleotide excision repair reconstituted with purified protein components," Cell **80**, 859–868.

ADAMS, J.M. and CORY, S. (1992). "Oncogene co-operation in leukaemogenesis," pages 119 to 141 in *Oncogenes in the Development of Leukaemia. Advances and Prospects in Clinical, Epidemiological and Laboratory Oncology: Cancer Surveys 15,* Witte, O.N., Ed. (Cold Spring Harbor Laboratory Press, Woodbury, New York).

ADAMS, L.M., ETHIER, S.P. and ULLRICH, R.L. (1987). "Enhanced *in vitro* proliferation and *in vivo* tumorigenic potential of mammary epithelium from BALB/c mice exposed *in vivo* to gamma-radiation and/or 7,12-dimethylbenz[a]anthracene," Cancer Res. **47**, 4425-4431.

ALBERTINE, K.H., STEINER, R.M., RADACK, D.M., GOLDING, D.M., PETERSON, D., COHN, H.E. and FARBER, J.L. (1998). "Analysis of cell type and radiographic presentation as predictors of the clinical course of patients with bronchoalveolar cell carcinoma," Chest **113**, 997–1006.

ALLEN, B.C., CRUMP, K.S. and SHIPP, A.M. (1988). "Correlation between carcinogenic potency of chemicals in animals and humans," Risk Anal. **8**, 531–544.

AMES, B.N., MAGAW, R. and GOLD, L.S. (1987). "Ranking possible carcinogenic hazards," Science **236**, 271–280.

ANDERSEN, A.C. and ROSENBLATT, L.S. (1969). "The effect of whole-body x-irradiation on the median lifespan of female dogs (beagles)," Radiat. Res. **39**, 177–200.

ANDERSON, S., AUQUIER, A., HAUCK, W.W., OAKES, D., VANDAELE, W. and WEISBERG, H.I. (1980). *Statistical Methods for Comparative Studies: Techniques for Bias Reduction* (John Wiley and Sons, Inc., Hoboken, New Jersey).

ANGELE, S. and HALL, J. (2000). "The ATM gene and breast cancer: Is it really a risk factor?" Mutat. Res. **462**, 167–178.

APPLEBY, J.M., BARBER, J.B., LEVINE, E., VARLEY, J.M., TAYLOR, A.M., STANKOVIC, T., HEIGHWAY, J., WARREN, C. and SCOTT, D.

(1997). "Absence of mutations in the ATM gene in breast cancer patients with severe responses to radiotherapy," Br. J. Cancer **76**, 1546–1549.

ARNISH, J. (1994). *Bayesian Estimation of Dose-Rate Effectiveness Factors*, M.S. Thesis, Department of Nuclear Engineering (University of Tennessee, Knoxville, Tennessee).

ARNISH, J. and GROER, P.G. (2000). "Bayesian estimation of dose rate effectiveness," Radiat. Prot. Dosim. **88**, 299–305.

ASZTERBAUM, M., EPSTEIN, J., ORO, A., DOUGLAS V., LEBOIT, P.E., SCOTT, M.P. and EPSTEIN, E.H., JR. (1999). "Ultraviolet and ionizing radiation enhance the growth of BCCs and trichoblastomas in patched heterozygous knockout mice," Nat. Med. **5**, 1285–1291.

AZZAM, E.I., DE TOLEDO, S.M. and LITTLE, J.B. (2004). "Stress signaling from irradiated to non-irradiated cells," Curr. Cancer Drug Targets **4**, 53–64.

BACHMANN, K. (1972). "Genome size in mammals," Chromosoma **37**, 85–93.

BAKALKIN, G., YAKOVLEVA, T., SELIVANOVA, G., MAGNUSSON, K.P., SZEKELY, L., KISELEVA, E., KLEIN, G., TERENIUS, L. and WIMAN, K.G. (1994). "*p53* binds single-stranded DNA ends and catalyzes DNA renaturation and strand transfer," Proc. Natl. Acad. Sci. USA **91**, 413–417.

BALMAIN, A. (2001). "Cancer genetics: From Boveri and Mendel to microarrays," Nat. Rev. Cancer **1**, 77–82.

BANDYOPADHYAY, S.K., D'ANDREA, E. and FLEISSNER, E. (1989). "Expression of cellular oncogenes: Unrearranged c-*myc* gene but altered promoter usage in radiation-induced thymoma," Oncogene Res. **4**, 311–318.

BARCELLOS-HOFF, M.H. and BROOKS, A.L. (2001). "Extracellular signaling through the microenvironment: A hypothesis relating carcinogenesis, bystander effects and genomic instability," Radiat. Res. **156**, 618–627.

BARENDSEN, G.W. (1978). "Fundamental aspects of cancer induction in relation to the effectiveness of small doses of radiation," pages 263 to 276 in *Late Biological Effects of Ionizing Radiation,* Proceedings Series, STI/PUB/489 (International Atomic Energy Agency, Vienna).

BARRETT, J.C. and WISEMAN, R.W. (1992). "Molecular carcinogenesis in humans and rodents," pages 1 to 30 in *Comparative Molecular Carcinogenesis, Progress in Clinical and Biological Research, Proceedings of the Fifth International Conference,* Vol. 376, Klein-Szanto, A.J., Anderson, M.W., Barrett, J.C. and Slaga, T.J., Eds. (Wiley-Liss, Hoboken, New Jersey).

BARTSCH, H., HOLLSTEIN, M., MUSTONEN, R., SCHMIDT, J., SPIETHOFF, A., WESCH, H., WIETHEGE, T. and MULLER, K.M. (1995). "Letter: Screening for putative radon-specific *p53* mutation hotspot in German uranium miners," Lancet **346**, 121.

BECKER, K.L. and GAZDAR, A.F. (1983). "The pulmonary endocrine cell and the tumors to which it gives rise," pages 161 to 188 in *Compara-*

tive Respiratory Tract Carcinogenesis, Reznik-Schuller, H.M., Ed. (CRC Press, Boca Raton, Florida).

BELINSKY, S.A. and ANDERSON, M.W. (1990). "Detection of non-*ras* oncogenes in 4-(methylnitrosamino)-1-(3-pyridyl)-1-butanone (NNK)-induced lung tumors from F344 rats," Proc. Amer. Assoc. Cancer Res. **31**, 778.

BELINSKY, S.A., FOLEY, J.F., WHITE, C.M., ANDERSON, M.W. and MARONPOT, R.R. (1990). "Dose-response relationship between ^{6}O-methylguanine formation in Clara cells and induction of pulmonary neoplasia in the rat by 4-(methylnitrosamino)-1-(3-pyridyl)-1-butanone," Cancer Res. **50**, 3772–3780.

BELINSKY, S.A., MIDDLETON, S.K., PICKSLEY, S.M., HAHN, F.F. and NIKULA, K.J. (1996). "Analysis of the K-*ras* and *p53* pathways in x-ray-induced lung tumors in the rat," Radiat. Res. **145**, 449–456.

BELINSKY, S.A., SWAFFORD, D.S., FINCH, G.L., MITCHELL C.E., KELLY G., HAHN, F.F., ANDERSON, M.W. and NIKULA, K.J. (1997). "Alterations in the K-*ras* and *p53* genes in rat lung tumors," Environ. Health Perspect. **105** (Suppl 4), 901–906.

BENNETT, J.M., CATOVSKY, D., DANIEL, M.T., FLANDRIN, G., GALTON, D.A., GRALNICK, H.R. and SULTAN, C. (1976). "Proposals for the classification of the acute leukaemias. French-American-British (FAB) co-operative group," Br. J. Haematol. **33**, 451–458.

BENSCH, K.G., CORRIN, B., PARIENTE, R. and SPENCER, H. (1968). "Oat-cell carcinoma of the lung. Its origin and relationship to bronchial carcinoid," Cancer **22**, 1163–1172.

BERG, J.W. (1996). "Morphological classification of human cancer," pages 28 to 44 in *Cancer Epidemiology and Prevention*, 2nd ed., Schottenfeld, D. and Fraumeni, J.F., Jr., Eds. (Oxford University Press, New York).

BERLIN, N.I. (1960). "An analysis of some radiation effects on mortality," pages 121 to 127 in *The Biology of Aging*, Strehler, B.L., Ebert, J.D., Glass, H.B. and Shock, N.W., Eds. (American Institute of Biological Sciences, Washington).

BERNARDO, J.M. and SMITH, A.F.M. (1994). *Bayesian Theory* (John Wiley and Sons, Inc., Hoboken, New Jersey).

BISSELL, M.J. and BARCELLOS-HOFF, M.H. (1987). "The influence of extracellular matrix on gene expression: Is structure the message?" J. Cell Sci. **8** (Suppl.), 327–343.

BLAIR, W.H. (1979). "The characterization of "oat cell-like" carcinoma of the lungs of rodents," (abstract) Proc. Am. Assoc. Cancer Res. **20**, 166.

BLUNT, T., FINNIE, N.J., TACCIOLI, G.E., SMITH, G.C., DEMENGEOT, J., GOTTLIEB, T.M., MIZUTA, R., VARGHESE, A.J., ALT, F.W., JEGGO, P.A. and JACKSON, S.P. (1995). "Defective DNA-dependent protein kinase activity is linked to V(D)J recombination and DNA repair defects associated with the murine *scid* mutation," Cell **80**, 813–823.

BOECKER, B.B., MUGGENBURG, B.A., MILLER, S.C. and BRADLEY, P.L., Eds. (1994). *Biennial Report on Long-Term Dose-Response Stud-*

ies of Inhaled or Injected Radionuclides 1 October 1991 – 30 September 1993 — Progress Report, U.S. Department of Energy Report ITRI-139 (National Technical Information Service, Springfield, Virginia).

BOECKER, B.B., GRIFFITH, W.C., HAHN, F.F., NIKULA, K.J., LUNDGREN, D.L. and MUGGENBURG, B.A. (1997). "Lifetime health risks from internally deposited, beta-emitting radionuclides," Radioprot. Coll. **32**, C1-343–C1-350.

BOND, V.P., SWIFT, M.N., TOBIAS, C.A. and BRECHER, G. (1952). "Bowel lesions following single deuteron irradiation," Fed. Proc. **11**, 408–409.

BONIVER, J., HUMBLET, C., RONGY, A.M., DELVENNE, C., DELVENNE, P., GREIMERS, R., THIRY, A., COURTOY, R. and DEFRESNE, M.P. (1990). "Cellular events in radiation-induced lymphomagenesis," Int. J. Radiat. Biol. **57**, 693–698.

BOS, J.L. (1989). "*ras* oncogenes in human cancer: A review," Cancer Res. **49**, 4682–4689.

BOUFFLER, S., SILVER, A., PAPWORTH, D., COATES, J. and COX, R. (1993). "Murine radiation leukaemogenesis: Relationship between interstitial teleomere-like sequences and chromosome 2 fragile sites," Genes Chromosomes Cancer **6**, 98–106.

BOUFFLER, S.D., MEIJNE, E.I.M., MORRIS, D.J. and PAPWORTH, D. (1997). "Chromosome 2 hypersensitivity and clonal development in murine radiation acute myeloid leukaemia," Int. J. Radiat. Biol. **72**, 181–189.

BOULTON, E., CLEARY, H. and PLUMB, M. (2002). "Myeloid, B and T lymphoid and mixed lineage thymic lumphomas in the irradiated mouse," Carcinogenesis **23**, 1079–1085.

BOUNACER, A., WICKER, R., CAILLOU, B., CAILLEUX, A.F., SARASIN, A., SCHLUMBERGER, M. and SUAREZ, H.G. (1997). "High prevalence of activating *ret* proto-oncogene rearrangements, in thyroid tumors from patients who had received external radiation," Oncogene 15, 1263–1273.

BRANNAN, C.I., PERKINS, A.S., VOGEL, K.S., RATNER, N., NORDLUND, M.L., REID, S.W., BUCHBERG, A.M., JENKINS, N.A., PARADA, L.F. and COPELAND, N.G. (1994). "Targeted disruption of the neurofibromatosis type-1 gene leads to developmental abnormalities in heart and various neural crest-derived tissues," Genes Dev. **8**, 1019–1029.

BRECHER, G., CRONKITE, E.P. and PEERS, J.H. (1953). "Neoplasms in rats protected against lethal doses of irradiation by parabiosis or para-aminopropriophenome," J. Natl. Cancer Inst. **14**, 159–175.

BRECKON, G., PAPWORTH, D. and COX, R. (1991). "Murine radiation myeloid leukaemogenesis: A possible role for radiation-sensitive sites on chromosome 2," Genes Chromosomes Cancer **3**, 367–375.

BREWEN, J.G. and BROCK, R.D. (1968). "The exchange hypothesis and chromosome-type aberrations," Mutat. Res. **6**, 245–255.

BRODEY, R.S., SAUER, R.M. and MEDWAY, W. (1963). "Canine bone neoplasms," J. Am. Vet. Med. Assoc. **143**, 471–495.

BRODY, S. (1924). "The kinetics of senescence," J. Gen. Physiol. **6**, 245–257.
BROOKS, A.L., GUILMETTE, R.A., EVANS, M.J. and DIEL, J.H. (1986). "The induction of chromosome aberrations in the livers of Chinese hamsters by injected thorotrast," pages 197 to 201 in *The Radiobiology of Radium and Thorotrast*, Gossner, W., Gerber, G.B., Hagen, U. and Luz, A., Eds. (Urban and Schwarzenberg, Baltimore).
BRUENGER, F.W., LLOYD, R.D., MILLER, S.C., TAYLOR, G.N., ANGUS, W. and HUTH, D.A. (1994). "Occurrence of mammary tumors in beagles given ^{226}Ra," Radiat. Res. **138**, 423–434.
BRUES, A.M. (1951). "Comparative chronic toxicities of radium and plutonium," pages 100 to 124 in *Argonne National Laboratory Quarterly Report*, U.S. Atomic Energy Commission Report ANL-4625 (National Technical Information Service, Washington).
BRUES, A.M. and SACHER, G.A. (1952). "Analysis of mammalian radiation injury and lethality," pages 441 to 465 in *Symposium on Radiobiology*, Nickson, J.J., Ed. (John Wiley and Sons, Inc., Hoboken, New Jersey).
BUCHMANN, A., BAUER-HOFMANN, R., MAHR, J., DRINKWATER, N.R., LUZ, A. and SCHWARZ, M. (1991). "Mutational activation of the c-Ha-*ras* gene in liver tumors of different rodent strains: Correlation with susceptibility to hepatocarcinogenesis," Proc. Natl. Acad. Sci. USA **88**, 911–915.
BUNN, P.A., JR., CHAN, D., DIENHART, D.G., TOLLEY, R., TAGAWA, M. and JEWETT, P.B. (1992). "Neuropeptide signal transduction in lung cancer: Clinical implications of bradykinin sensitivity and overall heterogeneity," Cancer Res. **52**, 24–31.
BUNZ, F., DUTRIAUX, A., LENGAUER, C., WALDMAN, T., ZHOU, S., BROWN, J.P., SEDIVY, J.M., KINZLER, K.W. and VOGELSTEIN, B. (1998). "Requirement for *p53* and *p21* to sustain G_2 arrest after DNA damage," Science **282**, 1497–1501.
BURKI, H.J. (1980). "Ionizing radiation-induced 6-thioguanine-resistant clones in synchronous CHO cells," Radiat. Res. **81**, 76–84.
BURNS, F.J. and ALBERT, R.E. (1986). "Radiation carcinogenesis in rat skin," pages 199 to 214 in *Radiation Carcinogenesis*, Upton, A.L., Albert, R.E., Burns, F.J., and Shore, T.E., Eds. (Elsevier Science, New York).
BURNS, F.J., SHORE, R.E., ROY, N., LOOMIS, C. and ZHAO, P. (2002). "*PTCH* (patched) and *XPA* genes in radiation-induced basal cell carcinomas," pages 175 to 178 in *Radiation and Homeostasis*, Sugahara, T., Nikaido, O. and Niwa, O., Eds. (Elsevier Science, New York).
CAHILL, D.P., KINZLER, K.W., VOGELSTEIN, B. and LENGAUER, C. (1999). "Genetic instability and Darwinian selection in tumours," Trends Cell Biol. **9**, M57–M60.
CARDIFF, R.D. and WELLINGS, S.R. (1999). "The comparative pathology of human and mouse mammary glands," J. Mammary Gland Biol. Neoplasia **4**, 105–122.

CARDIS. E. and ESTEVE, J. (1992). *International Collaborative Study of Cancer Risk Among Nuclear Industry Workers, I. Report of the Feasibility Study*, Internal Report No. 92/001 (International Agency for Research on Cancer, Lyon, France).

CARDIS, E., GILBERT, E.S., CARPENTER, L., HOWE, G., KATO, I., AMSTRONG, B.K., BERAL, V., COWPER, G., DOUGLAS, A., FIX J., FRY, S., KALDOR, J., LAVE, C., SALMON, L., SMITH, P., VOELZ, G. and WIGGS, L. (1995). "Effects of low doses and low dose rates of external ionizing radiation: Cancer mortality among nuclear industry workers in three countries," Radiat. Res. **142**, 117–132.

CARNES, B.A. and FRITZ, T.E. (1991). "Responses of the beagle to protracted irradiation. I. Effect of total dose and dose rate," Radiat. Res. **128**, 125–132.

CARNES, B.A. and OLSHANSKY, S.J. (1993). "Evolutionary perspectives on human senescence," Popul. Devel. Rev. **19**, 793–806.

CARNES, B.A. and OLSHANSKY, S.J. (1997). "A biologically motivated partitioning of mortality," Exp. Gerontol. **32**, 615–631.

CARNES, B.A., OLSHANSKY, S.J. and GRAHN, D. (1996). "Continuing the search for a law of mortality," Popul. Devel. Rev. **22**, 231–264.

CARNES, B.A., OLSHANSKY, S.J. and GRAHN, D. (1998). "An interspecies prediction of the risk of radiation-induced mortality," Radiat. Res. **149**, 487–492.

CARNES, B.A., GRAHN, D. and HOEL, D. (2003). "Mortality of atomic bomb survivors predicted from laboratory animals," Radiat. Res. **160**, 159–167.

CARR, Z.A., KLEINERMAN, R.A., STOVALL, M., WEINSTOCK, R.M., GRIEM, M.L. and LAND, C.E. (2002). "Malignant neoplasms after radiation therapy for peptic ulcer," Radiat. Res. **157,** 668–677.

CATTANEO, M.G., CODIGNOLA, A., VICENTINI, L.M., CLEMENTI, F. and SHER, E. (1993). "Nicotine stimulates a serotonergic autocrine loop in human small-cell lung carcinoma," Cancer Res. **53**, 5566–5568.

CHAKRABORTY, R. and SANKARANARAYANAN, K. (1995). "Cancer predisposition, radiosensitivity and the risk of radiation-induced cancers. II. A Mendelian single-locus model of cancer predisposition and radiosensitivity for predicting cancer risks in populations," Radiat. Res. **143**, 293–301.

CHALUT, C., MONCOLLIN, V. and EGLY, J. M. (1994). "Transcription by RNA polymerase II: A process linked to DNA repair," Bioessays **16**, 651–655.

CHARLES, D.R. (1950). "Radiation-induced mutations in mammals," Radiology **55**, 579–581.

CHEN, J., BIRKHOLTZ, G.G., LINDBLOM, P., RUBIO, C. and LINDBLOM, A. (1998). "The role of ataxia-telangiectasia heterozygotes in familial breast cancer," Cancer Res. **58**, 1376–1379.

CHEPKO, G. and SMITH, G.H. (1997). "Three division-component, structurally-distinct cell populations contribute to murine mammary epithelial renewal," Tissue Cell **29**, 239–253.

CHO, Y., GORINA, S., JEFFREY, P.D. and PAVLETICH, N.P. (1994). "Crystal structure of a *p53* tumor suppressor-DNA complex: Understanding tumorigenic mutations," Science **265**, 346–355.

CLARKE, A.R., PURDIE, C.A., HARRISON, D.J., MORRIS, R.G., BIRD, C.C., HOOPER, M.L. and WYLLIE, A.H. (1993). "Thymocyte apoptosis induced by *p53*-dependent and independent pathways," Nature **362**, 849–852.

CLAUS, H. (1976). "Genesis," pages X11 to X14 in *The Health Effects of Plutonium and Radium*, Jee, W.S.S., Ed. (The J.W. Press, Salt Lake City, Utah).

CLEARY, M.L. (1992). "Transcription factors in human leukaemias," pages 89 to 104 in *Oncogenes in the Development of Leukaemia. Advances and Prospects in Clinical, Epidemiological and Laboratory Oncology: Cancer Surveys 15*, Witte, O.N., Ed. (Cold Spring Harbor Laboratory Press, Woodbury, New York).

CLIFTON, K.H. (1979). "Animal models of breast cancer," pages 1 to 20 in *Endocrinology of Cancer*, Rose, D.P., Ed. (CRC Press, Boca Raton, Florida).

CLIFTON, K.H. (1986). "Thyroid cancer: Reevaluation of an experimental model for radiogenic endocrine carcinogenesis," pages 181 to 198 in *Radiation Carcinogenesis*, Upton, A.C., Albert, R.E., Burns, F.J., and Shore, R.E., Eds. (Elsevier Science, New York).

CLIFTON, K.H. (1990). "The clonogenic cells of the rat mammary and thyroid glands: Their biology, frequency of initiation and promotion/progression to cancer," pages 1 to 21 in *Scientific Issues in Quantitative Cancer Risk Assessment*, Moolgavkar, S.H., Ed. (Birkhauser, Boston).

CLIFTON, K.H. and CROWLEY, J.J. (1978). "Effects of radiation type and dose and the role of glucocorticoids, gonadectomy, and thyroidectomy in mammary tumor induction in mammotropin-secreting pituitary tumor-grafted rats," Cancer Res. **38**, 1507–1513.

CLIFTON, K.H. and FURTH, J. (1960). "Ducto-alveolar growth in mammary glands of adrenogonadectomized male rats bearing mammotropic pituitary tumors," Endocrinology **66**, 893–897.

CLIFTON, K.H. and SRIDHARAN, B.N. (1975). "Endocrine factors and tumor growth," pages 249 to 285 in *Cancer, A Comprehensive Treatise*, Becker, F.F., Ed. (Plenum Press, New York).

CLIFTON, K.H., DOUPLE, E.B. and SRIDHARAN, B.N. (1976). "Effects of grafts of single anterior pituitary glands on the incidence and type of mammary neoplasm in neutron- or gamma-irradiated Fischer female rats," Cancer Res. **36**, 3732–3735.

CLIFTON, K.H., YASUKAWA-BARNES, J., TANNER, M.A. and HANING, R.V., JR. (1985). "Irradiation and prolactin effects on rat mammary carcinogenesis: Intrasplenic pituitary and estrone capsule implants," J. Natl. Cancer Inst. **75**, 167–175.

COHEN, S.M. and ELLWEIN, L.B. (1990). "Cell proliferation in carcinogenesis," Science **249**, 1007–1011.

COLLINS, J.M., ZAHARKO, D.S., DEDRICK, R.L. and CHABNER, B.A. (1986). "Potential roles for preclinical pharmacology in phase I clinical trials," Cancer Treat. Rep. **70**, 73–80.

COLLINS, J.M., GRIESHABER, C.K. and CHABNER, B.A. (1990). "Pharmacologically guided phase I clinical trials based upon preclinical drug development," J. Natl. Cancer Inst. **82**, 1321–1326.

COMFORT, A. (1979). *The Biology of Senescence*, 3rd ed. (Elsevier Science, New York).

COQUERELLE, T.M., WEIBEZAHN, K.F. and LUCKE-HUHLE, C. (1987). "Rejoining of double strand breaks in normal human and ataxia-telangiectasia fibroblasts after exposure to ^{60}Co gamma-rays, ^{241}Am alpha-particles or bleomycin," Int. J. Radiat. Biol. **51**, 209–218.

CORNFORTH, M.N. (1990). "Testing the notion of the one-hit exchange," Radiat. Res. **121**, 21–27.

COROMINAS, M., SLOAN, S.R., LEON, J., KAMINO, H., NEWCOMB, E.W. and PELLICER, A. (1991). "*ras* activation in human tumors and in animal model systems," Environ. Health Perspect. **93**, 19–25.

COSGROVE, G.E., WALBURG, H.E. and UPTON, A.C. (1968). "Gastrointestinal lesions in aged conventional and germ free mice exposed to radiation as young adults," pages 303 to 312 in *Gastrointestinal Injury Excerpta Medica Foundation*, Sullivan, M.E., Ed. (Monogr. Nucl. Med. Biol., Amsterdam).

COX, D.R. (1972). "Regression models and life-tables," J. Royal Stat. Soc. **B34**, 187–220.

COX, R. (1982). "A cellular description of the repair defect in ataxia-telangiectasia," pages 141 to 153 in *Ataxia-Telangiectasia: A Cellular and Molecular Link Between Cancer, Neuropathology, and the Immune Deficiency*, Bridges, B.A. and Harnden, D., Eds. (John Wiley and Sons, Inc., Hoboken, New Jersey).

COX, D.R. and OAKES, D. (1984). *Analysis of Survival Data* (Chapman and Hall, New York).

COX, R., DEBENHAM, P.G., MASSON, W.K. and WEBB, M.B. (1986). "Ataxia-telangiectasia: A human mutation giving high-frequency misrepair of DNA double-stranded scissions," Mol. Biol. Med. **3**, 229–244.

CROSS, F.T. and MONCHAUX, G. (1999). "Risk assessment of radon health effects from experimental animal studies. A joint review of PNNL (USA) and CEA-COGEMA (France) data," pages 85 to 105 in *Indoor Radon Exposure and Its Health Consequences*, Inaba, J., Yonehara, H. and Doi, M., Eds. (Kodansha Scientific, Ltd., Tokyo).

CROUCH, E. and WILSON, R. (1979). "Interspecies comparison of carcinogenic potency," J. Toxicol. Environ. Health **5**, 1095–1118,

CROW, J.F. (1982). "How well can we assess genetic risks? Not very," pages 211 to 235 in *Critical Issues in Setting Radiation Dose Limits*, NCRP Proceedings No. 3 (National Council on Radiation Protection and Measurements, Bethesda, Maryland.

CRUMP, K., ALLEN, B. and SHIPP, A. (1989). "Choice of dose measure for extrapolating carcinogenic risk from animals to humans: An empirical investigation of 23 chemicals," Health Phys. **57** (Suppl. 1), 387–393.

CSH (1994). Cold Spring Harbor. *The Molecular Genetics of Cancer, Symposia on Quantitative Biology,* Vol. 59 (Cold Spring Harbor Laboratory Press, Woodbury, New York).

CURRAN, T. and VERMA, I.M. (1984). "FBR murine osteosarcoma virus. I. Molecular analysis and characterization of a 75,000-Da *gag*-fos fusion product," Virology **135**, 218–228.

CUTLER, R.G. and SEMSEI, I. (1989). "Development, cancer and aging: Possible common mechanisms of action and regulation," J. Gerontol. **44**, 25–34.

DAGLE, G.E., WELLER, R.E., FILIPY, R.E., WATSON, C.R. and BUSCHBOM, R.L. (1996). "The distribution and effects of inhaled ^{239}Pu(NO$_3$)$_4$ deposited in the liver of dogs," Health Phys. **71**, 198–205.

DAHLIN, D.C. (1978). *Bone Tumors: General Aspects and Data on 6,221 Cases,* 3rd ed. (Charles C. Thomas, Springfield, Illinois).

DALEY, G.Q., VAN ETTEN, R.A. and BALTIMORE, D. (1990). "Induction of chronic myelogenous leukemia in mice by the *p210 bcr/abl* gene of the Philadelphia chromosome," Science **247**, 824–830.

DALHEIMER, A.R., KAUL, A., SAID, M.D. and RIEDEL, W. (1986). "Investigation on the effect of incorporated radioactive and nonradioactive particles and their synergism by long-term animal studies – physiochemical and biokinetic properties of zirconium dioxide colloids," pages 167 to 171 in *The Radiobiology of Radium and Thorotrast,* Gossner, W., Gerber, G.B., Hagen, U. and Luz, A., Eds. (Urban and Schwarzenberg, Baltimore).

DEDRICK, R.L. and MORRISON, P.F. (1992). "Carcinogenic potency of alkylating agents in rodents and humans," Cancer Res. **52**, 2464–2467.

DEEVEY, E.S., JR. (1947). "Life tables for natural populations of animals," Q. Rev. Biol. **22**, 283–314.

DEGG, N.L., WEIL, M.M., EDWARDS, A., HAINES, J., COSTER, M., MOODY, J., ELLENDER, M., COX, R. and SILVER, A. (2003). "Adenoma multiplicity in irradiated *Apc(min)* mice is modified by chromosome 16 segments from Balb/c," Cancer Res. **63**, 2361–2363.

DEMPLE, B. and HARRISON, L. (1994). "Repair of oxidative damage to DNA: Enzymology and biology," Ann. Rev. Biochem. **63**, 915–948.

DENMAN, D.L., KIRCHNER, F.R. and OSBORNE, J.W. (1978). "Induction of colonic adenocarcinoma in the rat by x-irradiation," Cancer Res. **38**, 1899–1905.

DERINGER, M.K., HESTON, W.E. and LORENZ, E. (1954). "Effects of long-continued total body gamma irradiation on mice, guinea pigs and rabbits. IV. Actions on the breeding behavior of mice," pages 149 to 168 in *Biological Effects of External X and Gamma Radiation, Vol. I,* National Nuclear Energy Series, Zirkle, R.E., Ed. (McGraw-Hill Book Co., New York).

DEVEREUX, T.R., ANDERSON, M.W. and BELINSKY, S.A. (1991). "Role of *ras* protooncogene activation in the formation of spontaneous and nitrosamine-induced lung tumors in the resistant C3H mouse," Carcinogenesis **12**, 299–303.

DEVESA, S.S., SHAW, G.L. and BLOT, W.J. (1991). "Changing patterns of lung cancer incidence by histological type," Cancer Epidemiol. Biomarkers Prev. **1**, 29–34.

DIETRICH, W.F., LANDER, E.S., SMITH, J.S., MOSER, A.R., GOULD, K.A., LUONGO, C., BORENSTEIN, N. and DOVE, W. (1993). "Genetic identification of *Mom-1*, a major modifier locus affecting *min*-induced intestinal neoplasia in the mouse," Cell **75**, 631–639.

DJALALI, M., ADOLPH, S., STEINBACH, P., WINKING, H. and HAMEISTER, H. (1987). "A comparative mapping study of fragile sites in the human and murine genomes," Hum. Genet. **77**, 157–162.

DOMANN, F.E., MITCHEN, J.M. and CLIFTON, K.H. (1990). "Restoration of thyroid function after total thyroidectomy and quantitative thyroid cell transplantation," Endocrinology **127**, 2673–2678.

DONEHOWER, L.A., HARVEY, M., SLAGLE, B.L., MCARTHUR, M.J., MONTGOMERY, C.A., JR., BUTEL, J.S. and BRADLEY, A. (1992). "Mice deficient for *p53* are developmentally normal but susceptible to spontaneous tumours," Nature **356**, 215–221.

DONNER, E.M. and PRESTON, R.J. (1996). "The relationship between *p53* status, DNA repair and chromatid aberration induction in G2 mouse embryo fibroblast cells treated with bleomycin," Carcinogenesis **17**, 1161–1165.

DOUGHERTY, T.F., STOVER, B.J., DOUGHERTY, J.H., JEE, W.S.S., MAYS, C.W., REHFELD, C.E., CHRISTENSEN, W.R. and GOLDTHORPE, H.C. (1962). "Studies of the biological effects of Ra^{226}, Pu^{239}, Ra^{228} ($MsTh_1$), Th^{228} (RdTh) and Sr^{90} in adult beagles," Radiat. Res. **17**, 625–681.

DRAGANI, T.A. MANENTI, G. and PIEROTTI, M.A. (1995). "Genetics of murine lung tumors." Adv. Cancer Research **67**, 83–112.

DRAPKIN, R., REARDON, J.T., ANSARI, A., HUANG, J.C., ZAWEL, L., AHN, K., SANCAR, A. and REINBERG, D. (1994). "Dual role of TFIIH in DNA excision repair and in transcription by RNA polymerase II," Nature **368**, 769–772.

DRINKWATER, N.R., HANIGAN, M.H. and KEMP, C.J. (1989). "Genetic determinants of hepatocarcinogenesis in the $B6C3F_1$ mouse," Toxicol. Lett. **49**, 255–265.

DUHRSEN, U. and METCALF, D. (1990). "Effects of irradiation of recipient mice on the behavior and leukemogenic potential of factor-dependent hematopoietic cell lines," Blood **75**, 190–197.

DULBECCO, R., UNGER, M., ARMSTRONG, B., BOWMAN, M. and SYKA, P. (1983). "Epithelial cell types and their evolution in the rat mammary gland determined by immunological markers," Proc. Natl. Acad. Sci. USA **80**, 1033–1037.

DUMONT, J.E., MALONE, J.F. and VAN HERLE, A.J. (1980). *Irradiation and Thyroid Disease: Dosimetric, Clinical and Carcinogenic Aspects* (Commission of the European Communities, Brussels).

DUMOUCHEL, W. and GROER, P.G. (1989). "A Bayesian methodology for scaling radiation studies from animals to man," Health Phys. **57** (Suppl. 1), 411–418.

DUNCAN, A.M. and EVANS, H.J. (1983). "The exchange hypothesis for the formation of chromatid aberrations: An experimental test using bleomycin," Mutat. Res. **107**, 307–313.

DUNN, T.B. (1954). "Normal and pathologic anatomy of the reticular tissue in laboratory mice, with a classification and discussion of neoplasms," J. Nat. Cancer Inst. **14**, 1281–1433.

DUTTA, A., RUPPERT, J.M., ASTER, J.C. and WINCHESTER, E. (1993). "Inhibition of DNA replication factor RPA by *p53*," Nature **365**, 79–82.

DVIR, A., PETERSON, S.R., KNUTH, M.W., LU, H. and DYNAN, W.S. (1992). "Ku autoantigen is the regulatory component of a template-associated protein kinase that phosphorylates RNA polymerase II," Proc. Natl. Acad. Sci. USA **89**, 11920–11924.

EASTON, D.F. (1994). "Cancer risks in A-T heterozygotes," Int. J. Radiat. Biol. **66** (Suppl. 6), S177–S182.

EGGLESTON, J.C. (1984). "Bronchial carcinoids and their relationship to other pulmonary tumors with endocrine features," pages 389 to 405 in *The Endocrine Lung in Health and Disease*, Becker, K.L. and Gazdar, A.F., Eds. (Elsevier Science, New York).

EHLING, U.H. (1976). "Die Gefahrdung der menschlichen Erbanlagen im technischen Zeitalter," Fortschr. Rontgenstr. **124**, 166–171.

EHLING, U.H. (1991). "Mutations in the F_1 generation of mice," Prog. Clin. Biol. Res **372**, 481–496.

ELANDT-JOHNSON, R.C. and JOHNSON, N.L. (1980). *Survival Models and Data Analysis* (John Wiley and Sons, Inc., Hoboken, New Jersey).

ELEFANTY, A.G., HARIHARAN, I.K. and CORY, S. (1990). *"bcr/abl*, the hallmark of chronic myeloid leukaemia in man, induces multiple haematopoietic neoplasms in mice," EMBO J. **9**, 1069–1078.

EPA (2003). U.S. Environmental Protection Agency. *EPA Assessment of Risks from Radon in Homes*, Technical Report EPA 402-R-03-003 (National Technical Information Service, Springfield, Virginia).

ERFLE, V., SCHMIDT, J., STRAUSS, G.P., HEHLMANN, R. and LUZ, A. (1986). "Activation and biological properties of endogenous retroviruses in radiation osteosarcomagenesis," Leuk. Res. **10**, 905–913.

ETHIER, S.P., ADAMS, L.M. and ULLRICH, R.L. (1984). "Morphological and histological characteristics of mammary dysplasias occurring in cell dissociation-derived murine mammary outgrowth." Cancer Res. **44**, 4517–4522.

EVANS, R.D. (1943). "Protection of radium dial workers and radiologists from injury by radium," J. Indust. Hyg. **25**, 253–269.

EVANS, R.D. (1974). "Radium in man," Health Phys. **27**, 497–510.

EVANS, R.D., HARRIS, R.S. and BUNKER, J.W.M. (1944). "Radium metabolism in rats and the production of osteogenic sarcomas in experimental radium poisoning," Am. J. Roentgenol. Radium Ther. **52**, 353–373.

EVANS, R.D., KEANE, N.T., KOLENKOV, R.J., NEAL, W.R. and SHAHANAN, M.M. (1969). "Radiogenic tumors in the radium and mesothorium cases studied at M.I.T.," pages 157 to 194 in *Delayed Effects of Bone-Seeking Radionuclides*," Mays, C.W., Jee, W.S.S., Lloyd, R.D.,

Stover, B.J., Dougherty, J.H. and Taylor, G.N., Eds. (University of Utah Press, Salt Lake City, Utah).
FAILLA, G. (1958). "The aging process and cancerogenesis," Ann. NY Acad. Sci. **71**, 1124–1140.
FAILLA, G. (1960). "The aging process and somatic mutations," pages 170 to 175 in *The Biology of Aging*, Strehler, B.L., Ebert, J.D., Glass, H.B. and Shock, N.W., Eds. (American Institute of Biological Sciences, Washington).
FAILLA, G. and HENSHAW, P. (1931). "The relative biological effectiveness of x rays and gamma rays," Radiology **17**, 1–43.
FEARON, E.R. and VOGELSTEIN, B. (1990). "A genetic model for colorectal tumorigenesis," Cell **61**, 759–767.
FINKEL, M.P., REILLY, C.A., BISKIS, B.O. and GRECO, I.L. (1973). "Bone tumor viruses," pages 353 to 364 in *Bone: Certain Aspects of Neoplasia*, Price, C.H.G. and Ross, R.G.M., Eds. (Butterworths, London).
FINNIE, N.J., GOTTLIEB, T.M., BLUNT, T., JEGGO, P.A. and JACKSON, S.P. (1995). "DNA-dependent protein kinase activity is absent in *xrs-6* cells: Implications for site-specific recombination and DNA double-strand break repair," Proc. Natl. Acad. Sci. USA **92**, 320–324.
FINNON, R., MOODY, J., MEIJNE, E., HAINES, J., CLARK, D., EDWARDS, A., COX, R. and SILVER, A. (2002). "A major breakpoint cluster domain in murine radiation-induced acute myeloid leukemia," Mol. Carcinog. **34**, 64–71.
FISHEL, R. (2001). "The selection for mismatch repair defects in hereditary nonpolyposis colorectal cancer: Revising the mutator hypothesis," Cancer Res. **61**, 7369–7374.
FISHEL, R., LESCOE, M.K., RAO, M.R., COPELAND, N.G., JENKINS, N.A., GARBER, J., KANE, M. and KOLODNER, R. (1993). "The human mutator gene homolog MSH2 and its association with hereditary nonpolyposis colon cancer," Cell **75**, 1027–1038.
FORTINI, P., PASCUCCI, B., PARLANTI, E., D'ERRICO, M., SIMONELLI, V. and DOGLIOTTI, E. (2003). "The base excision repair: Mechanisms and its relevance for cancer susceptibility," Biochimie. **85**, 1053–1071.
FOX, M. (1975). "Factors affecting the quantitation of dose-response curves for mutation induction in V79 Chinese hamster cells after exposure to chemical and physical mutagens," Mutat. Res. **29**, 449–466.
FOX, J.C. and MCNALLY, N.J. (1988). "Cell survival and DNA double-strand break repair following x-ray or neutron irradiation of V79 cells," Int. J. Radiat. Biol. **54**, 1021–1030.
FRANKENBERG-SCHWAGER, M. (1990). "Induction, repair and biological relevance of radiation-induced DNA lesions in eukaryotic cells," Radiat. Environ. Biophys. **29**, 273–292.
FRANKENBERG-SCHWAGER, M. and FRANKENBERG, D. (1990). "DNA double-strand breaks: Their repair and relationship to cell killing in yeast," Int. J. Radiat. Biol. **58**, 569–575.

FRANKENBERG-SCHWAGER, M., FRANKENBERG, D. and HARBICH, R. (1994). "Radiation-induced mitotic gene conversion frequency in yeast is modulated by the conditions allowing DNA double-strand break repair," Mutat. Res. **314**, 57–66.

FRAZIER, M.E., SEED, T.M., WHITING, L.L. and STIEGLER, G.L. (1988). "Evidence for oncogene activation in radiation-induced carcinogenesis," pages 197 to 205 in *Multilevel Health Effects Research: From Molecules to Man, Proceedings of the 27th Hanford Symposium on Health the Environment*, Park, J.F. and Pelroy, R.A., Eds. (Battelle Press, Columbus, Ohio).

FREIREICH, E.J., GEHAN, E.A., RALL, D.P., SCHMIDT, L.H. and SKIPPER, H.E. (1966). "Quantitative comparison of toxicity of anticancer agents in mouse, rat, hamster, dog, monkey, and man," Cancer Chemother. Rep. **50**, 219–244.

FRIEBEN, A. (1992). "Demonstration eines Cancroids des rechten Handruckens, das sich nach langdauernder Einwirkung von Rontgenstrahlenr entwickelt hat Fortschr," Roentgenskr. **6**, 106–111.

FRIEDBERG, E.C., WALKER, G.C. and SIEDE, W. (1995). *DNA Repair and Mutagenesis* (ASM Press, Washington).

FRITZ, T.E., TOLLE, D.V. and SEED, T.M. (1985). "The preleukemic syndrome in radiation-induced myelogenous leukemia and related myeloproliferative disorders," pages 87 to 100 in *The Preleukemic Syndrome (Hemopoietic Dysplasia)*, Bagby, G.C., Jr., Ed. (CRC Press, Boca Raton, Florida).

FRY, R.J.M. (1980). "Extrapolation from experimental systems to man: A review of the problems and the possibilities," pages 111 to 136 in *To Address a Proposed Federal Radiation Research Agenda* (National Institutes of Health, Bethesda, Maryland).

FRY, S.A. (1998). "Studies of U.S. radium dial workers: An epidemiological classic," Radiat. Res. **150** (Suppl. 5), S21–S29.

FRY, R.J.M. and WITSCHI, H.P. (1984). "Lung tumors in mice," pages 63 to 78 in *Carcinogenesis and Mutagenesis Testing*, Douglas, J.F., Ed. (Humana Press, Clifton, New Jersey).

FRY, R.J.M., STORER, J.B. and BURNS, F.J. (1986). "Radiation induction of cancer of the skin," Br. J. Radiol. **19** (Suppl.), 58–60.

FUSE, E., KOBAYASHI, S., INABA, M., SUZUKI, H. and SUGIYAMA, Y. (1994). "Application of pharmacokinetically guided dose escalation with respect to cell cycle phase specificity" J. Natl. Cancer Inst. **86**, 989–996.

GAO, Y., CHAUDHURI, J., ZHU, C., DAVIDSON, L., WEAVER, D.T. and ALT, F.W. (1998). "A targeted DNA-PKcs-null mutation reveals DNA-PK-independent functions for KU in V(D)J recombination," Immunity **9**, 367–376.

GARD, D.L. and KROPF, D.L. (1993). "Confocal immunofluorescence microscopy of microtubules in oocytes, eggs, and embryos of algae and amphibians," Methods Cell Biol. **37**, 147–169.

GAZDAR, A.F. (1984). "The biology of endocrine tumors of the lung," pages 448 to 459 in *The Endocrine Lung in Health and Disease*, Becker, K.L. and Gazdar, A.F., Eds. (Elsevier Science, New York).
GEHAN, E.A. and SIDDIQUI, M.M. (1973). "Simple regression methods for survival time studies," J. Amer. Stat. Assoc. **68**, 848–856.
GERBER, G.B., WATSON, C.R., SUGAHARA, T. and OKADA, S. (1996). *International Radiobiology Archives of Long-Term Animal Studies, I. Descriptions of Participating Institutions and Studies*, U.S. Department of Energy Report DOE/RL-96-72 (National Technical Information Service, Springfield, Virginia).
GILBERT, E.S. (2001). "Invited Commentary: Studies of workers exposed to low doses of radiation," Am. J. Epidemiol. **153**, 319–322.
GILBERT, E.S., CROSS, F.T., SANDERS, C.L. and DAGLE, G.E. (1990). "Models for comparing lung-cancer risks in radon- and plutonium-exposed experimental animals," pages 783 to 802 in *Indoor Radon and Lung Cancer: Reality or Myth? Proceedings of the 29th Hanford Symposium on Health and the Environment*, Cross, F.T., Ed. (Battelle Press, Columbus, Ohio).
GILBERT, E.S., CROSS, F.T. and DAGLE, G.E. (1996). "Analysis of lung tumor risks in rats exposed to radon," Radiat. Res. **145**, 350–360.
GILBERT, E.S., GRIFFITH, W.C., BOECKER, B.B., DAGLE, G.E., GUILMETTE, R.A., HAHN, F.F., MUGGENBURG, B.A., PARK, J.F. and WATSON, C.R. (1998). "Statistical modeling of carcinogenic risks in dogs that inhaled $^{238}PuO_2$," Radiat. Res. **150**, 66–82.
GILBERT, E.S. KOSHURNIKOVA, N.A., SOKOLNIKOV, M., KHOKHRYAKOV, V.F., MILLER, S., PRESTON, D.L., ROMANOV, S.A., SHILNIKOVA, N.S., SUSLOVA, K.G. and VOSTROTIN, V.V. (2000). "Liver cancers in Mayak workers," Radiat. Res. **154**, 246–252.
GILBERT, E.S., STOVALL, M., GOSPODAROWICZ, M., VAN LEEUWEN, F.E., ANDERSSON, M., GLIMELIUS, B., JOENSUU, T., LYNCH, C.F., CURTIS, R.E., HOLOWATY, E., STORM, H., VAN'T VEER, M.B., FRAUMENI, J.F., JR., BOICE, J.D., JR., CLARKE, E.A. and TRAVIS, L.B. (2003). "Lung cancer after treatment for Hodgkin's disease: Focus on radiation effects," Radiat. Res. **159**, 161–173.
GISHIZKY, M.L. and WITTE, O.N. (1992). "Initiation of deregulated growth of multipotent progenitor cells by *bcr-abl in vitro*," Science **256**, 836–839.
GOEDECKE, W., PFEIFFER, P. and VIELMETTER, W. (1994). "Nonhomologous DNA end joining in *Schizosaccharomyces pombe* efficiently eliminates DNA double-strand-breaks from haploid sequences," Nucleic Acids Res. **22**, 2094–2101.
GOLDMAN, M., ROSENBLATT, L.S., HETHERINGTON, N.W. and FINKEL, M.P. (1973). "Scaling dose, time, and incidence of radium-induced osteosarcomas of mice and dogs to man," pages 347 to 357 in *Radionuclide Carcinogenesis, Proceedings of the Twelfth Annual Hanford Biology Symposium*, Sanders, C.L., Busch, R.H., Ballou, J.M. and Mahlum, D.D., Eds. (National Technical Information Service, Springfield, Virginia).

GOMPERTZ, B. (1825). "On the nature of the function expressive of the law of human mortality and on a new mode of determining the value of life contingencies," Phil. Trans. Roy. Soc. (London) **115**, 513–585.

GOODHEAD, D.T. (1994). "Initial events in the cellular effects of ionizing radiations: Clustered damage in DNA," Int. J. Radiat. Biol. **65**, 7–17.

GOSSNER, W. (1999). "Pathology of radium-induced bone tumors: New aspects of histopathology and histogenesis," Radiat. Res. **152** (Suppl. 6). S12–S15.

GOSSNER, W., HUG, O., LUZ, A. and MULLER, W.A. (1976). "Experimental induction of bone tumors by short-lived bone-seeking radionuclides," Recent Results Cancer Res. **54**, 36–49.

GOSSNER, W., GERBER, G.B., HAGEN, U. and LUZ, A., Eds. (1986). *The Radiobiology of Radium and Thorotrast* (Urban and Schwarzenberg, Baltimore).

GOTTLIEB, T.M. and JACKSON, S.P. (1993). "The DNA-dependent protein kinase: Requirement for DNA ends and association with Ku antigen," Cell **72**, 131–142.

GOTTLIEB, T.M. and JACKSON, S.P. (1994). "Protein kinases and DNA damage," Trends Biochem. Sci. **19**, 500–503.

GOWEN, L.C., JOHNSON, B.L., LATOUR, A.M., SULIK, K.K. and KOLLER, B.H. (1996). "*BRCA1* deficiency results in early embryonic lethality characterized by neuroepithelial abnormalities," Nat. Genet. **12**, 191–194.

GRAHAM, M.A. and WORKMAN, P. (1992). "The impact of pharmacokinetically guided dose escalation strategies in phase I clinical trials: Critical evaluation and recommendations for future studies," Ann. Oncol. **3**, 339–347.

GRAHN, D. (1959). "Genetic control of physiological processes: The genetics of radiation toxicity in animals" pages 181 to 190 in *Radioisotopes in the Biosphere*, Caldecott, R.S. and Snyder, L.A., Eds. (University of Minnesota, Minneapolis).

GRAHN, D. (1970). "Biological effects of protracted low dose radiation exposure of man and animals," pages 101 to 136 in *Late Effects of Radiation*, Fry, R.J.M., Grahn, D., Griem, M.L. and Rust, J.H., Eds. (Taylor and Francis, New York).

GRAHN, D. and SACHER, G. A. (1968). "Fractionation and protraction factors and the late effects of radiation in small mammals," pages 2-1 to 2-27 in *Dose Rate in Mammalian Biology*, Brown, D.G., Cragle, R.G. and Noonan, T. R., Eds., U.S. Atomic Energy Commission CONF-680410 (National Technical Information Service, Springfield, Virginia).

GRAHN, D., FRY, R.J.M. and LEA, R.A. (1972). "Analysis of survival and cause of death statistics for mice under single and duration-of-life gamma irradiation," Life Sci. Space Res. **10**, 175–186.

GRAHN, D., SACHER, G.A., LEA, R.A., FRY, R.J.M. and RUST, J.H. (1978). "Analytical approaches to and interpretations of data on time, rate, and cause of death of mice exposed to external gamma irradiation," pages 43 to 58 in *Late Biological Effects of Ionizing Radiation*,

Proceedings Series, STI/PUB/489 (International Atomic Energy Agency, Vienna).

GRAHN, D., LOMBARD, L.S. and CARNES, B.A. (1992). "The comparative tumorigenic effects of fission neutrons and cobalt-60 gamma rays in the B6CF$_1$ mouse," Radiat. Res. **129**, 19–36.

GRAHN, D., FOX, C., WRIGHT, B.J. and CARNES, B.A. (1994). *Studies of Acute and Chronic Radiation Injury at the Biological and Medical Research Division, Argonne National Laboratory, 1953-1970: Description of Individual Studies, Data Files, Codes, and Summaries of Significant Findings*, ANL-94/26 (Argonne National Laboratory, Argonne, Illinois).

GRAHN, D., WRIGHT, B.J., CARNES, B.A., WILLIAMSON, F.S. and FOX, C. (1995). *Studies of Acute and Chronic Radiation Injury at the Biological and Medical Research Division, Argonne National Laboratory, 1970-1992: The Janus Program: Survival and Pathology Data*, ANL-95/3 (Argonne National Laboratory Report, Argonne, Illinois).

GREENBERGER, J.S. (1991). "Toxic effects on the hematopoietic microenvironment," Exp. Hematol. **19**, 1101–1109.

GREENBERGER, J.S., WRIGHT, E., HENAULT, S., ANKLESARIA, P., LEIF, J., SAKAKEENY, M.A., FITZGERALD, T.J., PIERCE, J.H. and KASE, K. (1990). "Hematopoietic stem cell- and marrow stromal cell-specific requirements for gamma irradiation leukemogenesis *in vitro*," Exp. Hematol. **18**, 408–415.

GREENWOOD, M. (1928). "Laws of mortality from the biological point of view," J. Hyg. **28**, 267–294.

GRIFFIN, C.S., STEVENS, D.L. and SAVAGE, J.R. (1996). "Ultrasoft 1.5 keV aluminum K x rays are efficient producers of complex chromosome exchange aberrations as revealed by fluorescence *in situ* hybridization," Radiat. Res. **146**, 144–150.

GRIFFITH, W.C., BOECKER, B.B., WATSON, C.R. and GERBER, G.B. (1995). "Possible uses of animal databases for further statistical evaluation and modeling," pages 731 to 734 in *Radiation Research 1895 to 1995*, Hagen, U., Harder, D., Jung, H. and Streffer, C., Eds. (Universitatsdruckerei H. Sturtz AG, Wurzburg).

GROCH, K.M. and CLIFTON, K.H. (1992a). "The plateau phase rat goiter contains a sub-population of TSH-responsive follicular cells capable of proliferation following transplantation," Acta Endocrinol. (Copenh.) **126**, 85–96.

GROCH, K.M. and CLIFTON, K.H. (1992b). "The effects of goitrogenesis, involution, and goitrogenic rechallenge on the clonogenic cell content of the rat thyroid," Acta Endocrinol. (Copenh) **126**, 515–523.

GROMBACHER, T., MITRA, S. and KAINA, B. (1996). "Induction of the alkyltransferase (MGMT) gene by DNA damaging agents and the glucacorticoid dexamethasone and comparison with the response of base excision repair genes," Carcinogenesis **17**, 2329–2336.

GROSOVSKY, A.J. and LITTLE, J.B. (1985). "Evidence for linear response for the induction of mutations in human cells by x-ray exposures below 10 rads," Proc. Natl. Acad. Sci. USA **82**, 2092–2095.

GUERRERO, I., CALZADA, P., MAYER, A. and PELLICER, A. (1984). "A molecular approach to leukemogenesis: Mouse lymphomas contain an activated c-*ras* oncogene," Proc. Natl. Acad. Sci. USA **81**, 202–205.

GUILMETTE, R.A., GILLETT, N.A., EIDSON, A.F., GRIFFITH, W.C. and BROOKS, A.L. (1989). "The influence of non-uniform alpha irradiation of Chinese hamster liver on chromosome damage and the induction of cancer," pages 142 to 148 in *Risks from Radium and Thorotrast*, Taylor, D.M., Mays, C.W., Gerber, G.B. and Thomas, R.G., Eds. (British Institute of Radiology, London).

GUMBEL, E.J. (1954). *Statistical Theory of Extreme Value and Some Practical Applications: A Series of Lectures*, National Bureau of Standards Applied Mathematics Series No. 33 (U.S. Government Printing Office, Washington).

HAHN, W.C. and WEINBERG, R.A. (2002). "Rules for making human tumor cells," N. Engl. J. Med. **347**, 1593–1603.

HAHN, F.F., MUGGENBURG, B.A. and BOECKER, B.B. (1996). "Hepatic neoplasms from internally deposited $^{144}CeCl_3$," Toxicol. Path. **24**, 281–289.

HAHN, F.F., MUGGENBURG, B.A., MENACHE, M.G., GUILMETTE, R.A. and BOECKER, B.B. (1999). "Comparative stochastic effects of inhaled alpha- and beta-particle-emitting radionuclides in beagle dogs," Radiat. Res. **152**, (Suppl. 6), S19–S22.

HALEVY, O., RODEL, J., PELED, A. and OREN, M. (1991). "Frequent *p53* mutations in chemically induced murine fibrosarcoma," Oncogene **6**, 1593–1600.

HALL, E.J. (2000). *Radiobiology for the Radiologist*, 5th ed. (J.B. Lippincott Company, Philadelphia).

HALL, E.J., SCHIFF, P.B., HANKS, G.E., BRENNER, D.J., RUSSO, J., CHEN, J., SAWANT, S.G. and PANDITA, T.K. (1998). "A preliminary report: Frequency of A-T heterozygotes among prostate cancer patients with severe late responses to radiation therapy," Cancer J. Sci. Am. **4**, 385–389.

HANAHAN, D. and WEINBERG, R.A. (2000). "The hallmarks of cancer," Cell **100**, 57–70.

HARE, W.V. and STEWART, H.L. (1956). "Chronic gastritis of the glandular stomach, adenomatous polyps of the duodenum, and calcareous pericarditis in strain DBA mice," J. Nat. Cancer Inst. **16**, 889–911.

HARIHARAN, I.K., HARRIS, A.W., CRAWFORD, M., ABUD, H., WEBB, E., CORY, S. and ADAMS, J.M. (1989). "A *bcr-v-abl* oncogene induces lymphomas in transgenic mice," Mol. Cell Biol. **9**, 2798–2805.

HARPER, J.W., ADAMI, G.R., WEI, N., KEYOMARSI, K. and ELLEDGE, S.J. (1993). "The *p21* Cdk-interacting protein Cip1 is a potent inhibitor of G1 cyclin-dependent kinases," Cell **75**, 805–816.

HARRIS, C.C. (1996). "*p53* tumor suppressor gene: From the basic research laboratory to the clinic—an abridged historical perspective," Carcinogenesis **17**, 1187–1198.

HARRIS, L.C., POTTER, P.M., TANO, K., SHIOTA, S., MITRA, S. and BRENT, T.P. (1991). "Characterization of the promoter region of the human O6-methylguanine-DNA methyltransferase gene," Nucleic Acids Res. **19**, 6163–6167.

HART, R.W. and SETLOW, R.B. (1975). "DNA repair and life span of animals," pages 801 to 804 in *Molecular Mechanisms for Repair of DNA, Part B*, Hanawalt, P.C. and Setlow, R.B., Eds. (Plenum Press, New York).

HARVEY, M., MCARTHUR, M.J., MONTGOMERY, C.A., JR., BUTEL, J.S., BRADLEY, A. and DONEHOWER, L.A. (1993). "Spontaneous and carcinogen-induced tumorigenesis in *p53*-deficient mice," Nat. Genet. **5**, 225–229.

HAUPT, Y., HARRIS, A.W. and ADAMS, J.M. (1992). "Retroviral infection accelerates T lymphomagenesis in E mu-N-*ras* transgenic mice by activating c-*myc* and N-*myc*," Oncogene **7**, 981–986.

HAYATA, I. (1983). "Partial deletion of chromosome 2 in radiation-induced myeloid leukemia in mice," pages 277 to 297 in *Radiation-Induced Chromosome Damage in Man*, Ishihara, T. and Sasaki, M.S., Eds. (A.R. Liss, New York).

HAYATA, I., SEKI, M., YOSHIDA, K., HIRASHIMA, K., SADO, T., YAMAGIWA, J. and ISHIHARA, T. (1983). "Chromosomal aberrations observed in 52 mouse myeloid leukemias," Cancer Res. **43**, 367–373.

HAZARD, J.B., HAWK, W.A. and CRILE, G. (1959). "Medullary (solid) carcinoma of the thyroid: A clinicopathologic entity," J. Clin. Endocrinol. Metab. **19**, 152–161.

HEARTLEIN, M.W. and PRESTON, R.J. (1985). "An explanation of interspecific differences in sensitivity to x-ray-induced chromosome aberrations and a consideration of dose-response curves," Mutat. Res. **150**, 299–305.

HEIDENREICH, W.F., COLLIER, C., MORLIER, J.P., CROSS, F.T., KAISER, J.C. and MONCHAUX, G. (2004). "Age-adjustment in experimental animal data and its application to lung cancer in radon-exposed rats," Radiat. Environ. Biophys. **43**, 183–188.

HEISTERKAMP, N., JENSTER, G., TEN HOEVE, J., ZOVICH, D., PATTENGALE, P.K. and GROFFEN, J. (1990). "Acute leukaemia in *bcr/abl* transgenic mice," Nature **344**, 251–253.

HENDERSON, B.E., ROSS, R. and BERNSTEIN, L. (1988). "Estrogens as a cause of human cancer: The Richard and Hinda Rosenthal Foundation Award Lecture," Cancer Res. **48**, 246–253.

HERRANZ, M., SANTOS, J., SALIDO, E., FERNANDEZ-PIQUERAS, J. and SERRANO, M. (1999). "Mouse *p73* gene maps to the distal part of chromosome 4 and might be involved in the progression of gamma-radiation-induced T-cell lymphomas," Cancer Res. **59**, 2068–2071.

HINO, O., KLEIN-SZANTO, A.J.P., FREED, J.J., TESTA, J.R., BROWN, D.Q., VILENSKY, M., YEUNG, R.S., TARTOF, K.D. and KNUDSON, A.G. (1993). "Spontaneous and radiation-induced renal tumors in the Eker rat model of dominantly inherited cancer," Proc. Natl. Acad. Sci. USA **90**, 327–331.

HOEL, D.G. (1972). "A representation of mortality data by competing risks," Biometics **28**, 475–488.
HOLLAND, J.M. and FRY, R.J.M. (1982). "Neoplasms of the integumentary system and Harderian gland," pages 513 to 528 in *The Mouse in Biomedical Research*, Foster, H.L., Small, J.D. and Fox, J.G., Eds. (Academic Press, New York).
HOLM, L.E., WIKLUND, K.E., LUNDELL, G.E., BERGMAN, N.A., BJELKENGREN, G., CEDERQUIST, E.S., ERICSSON U.BC., LARSSON, L.G, LINDBERG, M.E., LINDBERG, R.S., WICKLUND, H.V. and BOICE, J.D., JR. (1988). "Thyroid cancer after diagnostic doses of iodine-131: A retrospective cohort study," J. Natl. Cancer Inst. **80**, 1132–1138.
HOWELL, R.W., WESSELS, B.W., LOEVINGER, R., WATSON, E.E., BOLCH, W.E., BRILL, A.B., CHARKES, N.D., FISHER, D.R., HAYS, M.T., ROBERTSON, J.S., SIEGEL, J.A. and THOMAS, S.R. (1999). "The MIRD perspective 1999. Medical International Radiation Dose Committee," J. Nucl. Med. **40**, 3S–10S.
HUBBARD, K., IDE, H., ERLANGER, B.F. and WALLACE, S.S. (1989). "Characterization of antibodies to dihydrothymine, a radiolysis product of DNA," Biochemistry **28**, 4382–4387.
HUGGINS, C., GRAND, L.C. and BRILLANTES, F.P. (1959). "Critical significance of breast structure in the induction of mammary cancer in the rat," Proc. Natl. Acad. Sci. USA **45**, 1294–1300.
HULSE, E.V. (1962). "Tumours of the skin and other delayed effects of external beta irradiation of mice using ^{90}Sr and ^{32}P," Brit. J. Cancer **16**, 72–86.
HUNTER, T. (1991). "Cooperation between oncogenes," Cell **64**, 249–270.
IARC (1994). International Agency for Research on Cancer. *Pathology of Tumours in Laboratory Animals, Volume II, Tumours of the Mouse*, IARC Scientific Publication No. 111, 2nd ed., Turusov, V.S. and Mohr, U., Eds. (International Agency for Research on Cancer, Lyon, France).
IARC (2000), International Agency for Research on Cancer. *Ionizing Radiation, Part I: X- and Gamma-Radiation and Neutrons*, IARC Monographs Vol. 75 (International Agency for Research on Cancer, Lyon, France).
IARC (2001). International Agency for Research on Cancer. *Ionizing Radiation, Part II: Some Internally Deposited Radionuclides*, IARC Monographs Vol. 78 (International Agency for Research on Cancer Press, Lyon, France).
ICRP (1979). International Commission on Radiological Protection. "Dosimetric model for bone," pages 35 to 46 in *Limits for Intake of Radionuclides by Workers*, ICRP Publication 30, Part 1, Ann. ICRP **2**(3/4) (Elsevier Science, New York).
ICRP (1989). International Commission on Radiological Protection. *Age-Dependent Doses to Members of the Public from Intake of Radionuclides: Part 1*, ICRP Publication 56, Ann. ICRP **20**(2) (Elsevier Science, New York).

ICRP (1991). International Commission on Radiological Protection. *1990 Recommendations of the International Commission on Radiological Protection*, ICRP Publication 60, Ann. ICRP **21**(1–3) (Elsevier Science, New York).

ICRP (1993). International Commission on Radiological Protection. *Age-Dependent Doses to Members of the Public from Intake of Radionuclides: Part 2, Ingestion Dose Coefficients*, ICRP Publication 67, Ann. ICRP **23**(3/4) (Elsevier Science, New York).

ICRP (1995a). International Commission on Radiological Protection. *Age-Dependent Doses to Members of the Public from Intake of Radionuclides: Part 3, Ingestion Dose Coefficients*, ICRP Publication 69, Ann. ICRP **25**(1) (Elsevier Science, New York).

ICRP (1995b). International Commission on Radiological Protection. *Age-Dependent Doses to Members of the Public from Intake of Radionuclides: Part 4, Inhalation Dose Coefficients*, ICRP Publication 71, Ann. ICRP **25**(3–4) (Elsevier Science, New York).

ICRP (1999). International Commission on Radiological Protection. *Genetic Susceptibility to Cancer*, ICRP Publication 79, Ann. ICRP **28**(1–2) (Elsevier Science, New York).

ICRU (2002). International Commission on Radiation Units and Measurements. *Absorbed Dose Specification in Nuclear Medicine*, ICRU Report No. 67 (International Commission on Radiation Units and Measurements, Bethesda, Maryland).

IMAMURA, N., ABE, K. and OGUMA, N. (2002). "High incidence of point mutations of *p53* suppressor oncogene in patients with myelodysplastic syndrome among atomic-bomb survivors: A 10-year follow-up," Leukemia **16**, 154–156.

INSKIP, P.D., STOVALL, M. and FLANNERY, J.T. (1994). "Lung cancer risk and radiation dose among women treated for breast cancer," J. Natl. Cancer Inst. **86**, 983–988.

JACKS, T., FAZELI, A., SCHMITT, E.M., BRONSON, R.T., GOODELL, M.A. and WEINBERG, R.A. (1992). "Effects of an *Rb* mutation in the mouse," Nature **359**, 295–300.

JACKSON, S.P. (2002). "Sensing and repairing DNA double-strand breaks," Carcinogenesis **23**, 687–696.

JACKSON-GRUSBY, L. (2002). "Modeling cancer in mice," Oncogene **21**, 5504–5514.

JANOWSKI, M., COX, R. and STRAUSS, P.G. (1990). "The molecular biology of radiation-induced carcinogenesis: Thymic lymphoma, myeloid leukaemia and osteosarcoma," Int. J. Radiat. Biol. **57**, 677–691.

JEGGO, P.A., TACCIOLI, G.E. and JACKSON, S.P. (1995). "Menage a trois: Double strand break repair, V(D)J recombination and DNA-PK," Bioessays **17**, 949–957.

JHIANG, S.M., CHO, J.Y., FURMINGER, T.L., SAGARTZ, J.E., TONG, Q., CAPEN, C.C. and MAZZAFERRI, E.L. (1998). "Thyroid carcinomas in *RET/PTC* transgenic mice," Recent Results Cancer Res. **154**, 265–270.

JOHNSON, R.D., LIU, N. and JASIN, M. (1999). "Mammalian *XRCC2* promotes the repair of DNA double-strand breaks by homologous recombination," Nature **401**, 397–399.
JOISHY, S.K., COOPER, R.A. and ROWLEY, P.T. (1977). "Alveolar cell carcinoma in identical twins. Similarity in time of onset, histochemistry, and site of metastasis," Ann. Intern. Med. **87**, 447–450.
JOSTES, R.F., BUSHNELL, K.M. and DEWEY, W.C. (1980). "X-ray induction of 9-azoguanine-resistant mutants in synchronous Chinese hamster ovary cells," Radiat. Res. **83**, 146–161.
KAISER, J.C., HEIDENREICH, W.F., MONCHAUX, G., MORLIER, J.P. and COLLIER, C.G. (2004). "Lung tumour risk in radon-exposed rats from different experiments: Comparative analysis with biologically based models," Radiat. Environ. Biophys. **43**, 189–201.
KALBFLEISCH, J.D. and PRENTICE, R.L. (1980). *The Statistical Analysis of Failure Time Data* (John Wiley and Sons, Inc., Hoboken, New Jersey).
KALDOR, J.M., DAY, N.E. and HEMMINKI, K. (1988). "Quantifying the carcinogenicity of antineoplastic drugs," Eur. J. Cancer Clin. Oncol. **24**, 703–711.
KALMAN, E. (1970). "Is it possible to utilize an experimental osteosarcoma in clinical practice?" Bratisl. Lek. Listy **54**, 401–406.
KAMADA, N., TANAKA, K., OGUMA, N. and MABUCHI, K. (1991). "Cytogenetic and molecular changes in leukemia among atomic bomb survivors," J. Radiat. Res. **32** (Suppl. 2), 257–265.
KAMIYA, K., GOULD, M.N. and CLIFTON, K.H. (1991). "Differential control of alveolar and ductal development in grafts of monodispersed rat mammary epithelium," Proc. Soc. Exp. Biol. Med. **196**, 284–292.
KAMIYA, K., GOULD, M.N. and CLIFTON, K.H. (1998). "Quantitative studies of ductal versus alveolar differentiation from mammary clonogens," Proc. Soc. Exp. Biol. Med. **219**, 217–225.
KANAAR, R., HOEIJMAKERS, J.H. and VAN GENT, D.C. (1998). "Molecular mechanisms of DNA double strand break repair," Trends Cell Biol. **8**, 483–489.
KAPLAN, H.S. (1974). "Leukemia and lymphoma in animals," pages 94 to 163 in *The Etiology of Leukemia, Ser. Haematol.*, Jensen, K.G. and Killimann, S.A., Eds. (Munksgaard, Copenhagen).
KAPLAN, E.L. and MEIER, P. (1958). "Nonparametric estimation from incomplete observations," J. Am. Stat. Assoc. 53, 457–481.
KAZANTSEV, A. and SANCAR, A. (1995). "Does the *p53* up-regulated Gadd45 protein have a role in excision repair?" Science **270**, 1003–1004.
KEEHN, R., AUERBACH, O., NAMBU, S., CARTER, D., SHIMOSATO, Y., GREENBERG, S.D., TATEISHI, R., SACCOMANNO, G., TOKUOKA, S. and LAND, C. (1994). "Reproducibility of major diagnoses in a binational study of lung cancer in uranium miners and atomic bomb survivors," Am. J. Clin. Pathol. **101**, 478–482.
KEITH, C.T. and SCHREIBER, S.L. (1995). "PIK-related kinases: DNA repair, recombination, and cell cycle checkpoints," Science **270**, 50–51.

KELLOFF, G.J., LANE, W.T., TURNER, H.C. and HUEBNER, R.J. (1969). "*In vivo* studies of the FBJ murine osteosarcoma virus," Nature **223**, 1379–1380.

KELLY, K., KANE, M.A. and BUNN, P.A., JR. (1991). "Growth factors in lung cancer: Possible etiologic role and clinical target," Med. Pediatr. Oncol. **19**, 450–458.

KERN, S.E., KINZLER, K.W., BRUSKIN, A., JAROSZ, D., FRIEDMAN, P., PRIVES, C. and VOGELSTEIN, B. (1991). "Identification of *p53* as a sequence-specific DNA-binding protein," Science **252**, 1708–1711.

KHANNA, K.K., LAVIN, M.F., JACKSON, S.P. and MULHERN, T.D. (2001). "ATM, a central controller of cellular responses to DNA damage," Cell Death Differ. **8**, 1052–1065.

KIM, N.D. and CLIFTON, K.H. (1993). "Characterization of rat mammary epithelial cell subpopulations by peanut lectin and anti-Thy-1.1 antibody and study of flow-sorted cells *in vivo*," Exp. Cell Res. **207**, 74–85.

KIM, U., FURTH, J. and CLIFTON, K.H. (1960). "Relation of mammary tumors to mammotropes. III. Hormone responsiveness of transplanted mammary tumors," Proc. Soc. Exp. Biol. Med. **103**, 646–650.

KIM, N.D., OBERLEY, T.D. and CLIFTON, K.H. (1993). "Primary culture of flow cytometry-sorted rat mammary epithelial cell (RMEC) subpopulations in a reconstituted basement membrane, Matrigel," Exp. Cell. Res. **209**, 6–20.

KIRKWOOD, T.B. (1992). "Comparative life spans of species: Why do species have the life spans they do?" Am. J. Clin. Nutr. **55** (Suppl. 6), 1191S–1195S.

KIRSTEN, W.H., ANDERSON, D.G., PLATZ, C.E. and CROWElL, E.B., JR. (1962). "Observations on the morphology and frequency of polyoma tumors in rats," Cancer Res. **22**, 484–491.

KLEINBAUM, D.G. (2005). *Survival Analysis: A Self-Learning Text*, 2nd ed. (Springer, New York).

KORDON, E.C. and SMITH, G.H. (1998). "An entire functional mammary glad may comprise the progeny from a single cell," Development **125**, 1921–1930.

KOSHURNIKOVA, N.A., SHILNIKOVA, N.S., OKATENKO, P.V., KRESLOV, V.V., BOLOTNIKOVA, M.G., SOKOLNIKOV, M.E., KHOKRIAKOV, V.F., SUSLOVA, K.G., VASSILENKO, E.K. and ROMANOV, S.A. (1999). "Characteristics of the cohort of workers at the Mayak nuclear complex," Radiat. Res. **152**, 352–363.

KOSHURNIKOVA, N.A., GILBERT, E.S., SOKOLNIKOV, M., KHOKHRYAKOV, V.F., MILLER, S., PRESTON, D.L., ROMANOV, S.A., SHILNIKOVA, N.S., SUSLOVA, K.G. and VOSTROTIN, V.V. (2000). "Bone cancers in Mayak workers," Radiat. Res. **154**, 237–245.

KOSHURNIKOVA, N.A., GILBERT, E.S., SHILNIKOVA, N.S., SOKOLNIKOV, M., PRESTON, D.L., KREISHEIMER, M., RON, E., OKATENKO, P. and ROMANOV, S.A. (2002). "Studies on the Mayak nuclear workers: Health effects," Radiat. Environ. Biophys. **41**, 29–31.

KREISHEIMER, M., KOSHURNIKOVA, N.A., NEKOLLA, E., KHOKHRYAKOV, V.F., ROMANOV, S.A., SOKOLNIKOV, M.E.,

SHILNIKOVA, N.S., OKATENKO, P.V. and KELLERER, A.M. (2000). "Lung cancer mortality among male nuclear workers of the Mayak facilities in the former Soviet Union," Radiat. Res. **154**, 3–11.

KREISHEIMER, M., SOKOLNIKOV, M.E., KOSHURNIKOVA, N.A., KHOKHRYAKOV, V.F., ROMANOV, S.A., SHILNIKOVA, HN.S., OKATENKO, P.V., NEKOLLA, E., and KELLERER, A.M. (2003). "Lung cancer mortality among nuclear workers of the Mayak facilities in the former Soviet Union: An updated analysis considering smoking as the main confounding factor," Radiat. Environ. Biophys. **42**, 129–135.

KUMAR, R., SUKUMAR, S. and BARBACID, M. (1990). "Activation of *ras* oncogenes preceding the onset of neoplasia," Science **248**, 1101–1104.

KUSEWITT, D.F., APPLEGATE, L.A. and LEY, R.D. (1991). "Ultraviolet radiation-induced skin tumors in a South American opossum (*Monodelphis domestica*)," Vet. Pathol. **28**, 55–65.

LAND, C.E. and SINCLAIR, W.K. (1991). "The relative contributions of different organ sites to the total cancer mortality associated with low-dose radiation exposure," pages 31 to 57 in *Risks Associated with Ionising Radiations*, Ann. ICRP **22**(1) (Elsevier Science, New York).

LAND, C.E., SHIMOSATO, Y., SACCOMANNO, G., TOKUOKA, S., AUERBACH, O., TATEISHI, R., GREENBERG, S.D., NAMBU, S., CARTER, D., AKIBA, S., KEEHN, R., MADIGAN, P., MASON, T.J. and TOKUNAGA, M. (1993a). "Radiation-associated lung cancer: A comparison of the histology of lung cancers in uranium miners and survivors of the atomic bombings of Hiroshima and Nagasaki," Radiat. Res. **134**, 234–243.

LAND, C.E., TOKUNAGA, M., TOKUOKA, S. and NAKAMURA, N. (1993b). "Letter: Early-onset breast cancer in A-bomb survivors," Lancet **342**, 237.

LANGDON, W.Y., HARRIS, A.W., CORY, S. and ADAMS, J.M. (1986). "The c-*myc* oncogene perturbs B lymphocyte development in E-mu-*myc* transgenic mice," Cell **47**, 11–18.

LANGHAM, W.H. and HEALY, J.W. (1973). "Maximum permissible body burdens and concentration of plutonium: Biological basis and history of development," pages 569 to 592 in *Uranium, Plutonium, Transplutonic Elements*, Hodge, H.C., Stannard, J.N. and Hursh, J.B., Eds. (Springer-Verlag, New York).

LAVIN, M. (1998). "Role of the ataxia-telangiectasia gene (ATM) in breast cancer. A-T heterozygotes seem to have an increased risk but its size is unknown," Brit. Med. J. **317**, 486–487.

LAWLESS, J.F. (2002). *Statistical Models and Methods for Lifetime Data*, 2nd ed. (John Wiley and Sons, Inc., Hoboken, New Jersey).

LE, X.C., XING, J.Z., LEE, J., LEADON, S.A. and WEINFELD, M. (1998). "Inducible repair of thymine glycol detected by an ultrasensitive assay for DNA damage," Science **280**, 1066–1069.

LEACH, F.S., NICOLAIDES, N.C., PAPADOPOULOS, N., LIU, B., JEN, J., PARSONS, R., PELTOMAKI, P., SISTONEN, P., AALTONEN, L.A., NYSTROM-LAHTI, M., GUAN, X.Y, ZHANG, J., MELTZER, P.S., YU,

J.W, KAO, F.T, CHEN, D.J., CEROSALETTI, K.M., FOURNIER, R.E.K., TODD, S., LEWIS, T., LEACH, R.J., NAYLOR, S.L., WEISSENBACH, J., MECKLIN, J.P, JARVINEN, H., PETERSEN, G.M., HAMILTON, S.R., GREEN, J., JASS, J., WATSON, P., LYNCH, H.T., TRENT, J.M., DE LA CHAPELLE, A., KINZLER, K.W. and VOGELSTEIN, B. (1993). "Mutations of a *mutS* homolog in hereditary nonpolyposis colorectal cancer," Cell **75**, 1215–1225.

LEADON, S.A. and HANAWALT, P.C. (1983). "Monoclonal antibody to DNA containing thymine glycol," Mutat. Res. **112**, 191–200.

LE BEAU, M.M. (1992). "Deletions of chromosome 5 in malignant myeloid disorders," pages 143 to 159 in *Oncogenes in the Development of Leukaemia. Advances and Prospects in Clinical, Epidemiological and Laboratory Oncology: Cancer Surveys 15,* Witte, O.N., Ed. (Cold Spring Harbor Laboratory Press, Woodbury, New York).

LEE, J.M. and BERNSTEIN, A. (1993). "*p53* mutations increase resistance to ionizing radiation," Proc. Natl. Acad. Sci. USA **90**, 5742–5746.

LEE, W., CHIACCHIERINI, R.P., SCHLEIN, B. and TELLES, N.C. (1982). "Thyroid tumors following ^{131}I or localized x irradiation to the thyroid and pituitary glands in rats," Radiat. Res. **92**, 307–319.

LEHNERT, B.E. and IYER, R. (2002). "Exposure to low-level chemicals and ionizing radiation: Reactive oxygen species and cellular pathways," Hum. Exp. Toxicol. **21**, 65–69.

LEMOINE, N.R., MAYALL, E.S., WILLIAMS, E.D., THURSTON, V. and WYNFORD-THOMAS, D. (1988). "Agent-specific *ras* oncogene activation in rat thyroid tumors," Oncogene **3**, 541–544.

LEVY, D.B., SMITH, K.J., BEAZER-BARCLAY, Y., HAMILTON, S.R., VOGELSTEIN, B. and KINZLER, K.W. (1994). "Inactivation of both *APC* alleles in human and mouse tumors," Cancer Res. **54**, 5953–5958.

LI, R., WAGA, S., HANNON, G.J., BEACH, D. and STILLMAN, B. (1994). "Differential effects by the *p21* CDK inhibitor on PCNA-dependent DNA replication and repair," Nature **371**, 534–537.

LICHTENSTEIN, L. (1977). *Bone Tumors*, 5th ed. (C.V. Mosby, St. Louis, Missouri).

LINNOILA, I. (1990). "Pathology of non-small cell lung cancer. New diagnostic approaches," Hematol. Oncol. Clin. North Am. **4**, 1027–1051.

LISCO, H. and FINKEL, M.P. (1946). *Carcinogenic Action of Some Substances which May be a Problem in Certain Future Industries*, Document MDCC-145 (Technical Information Center, Oak Ridge, Tennessee).

LITTLE, J.B. (1998). "Radiation-induced genomic instability," Int. J. Radiat. Biol. **74**, 663–671.

LITTLE, J.B. (2003). "Genomic instability and bystander effects: A historical perspective," Oncogene **22**, 6978–6987.

LITVINOV, N.N. and SOLOVIEV, J.N. (1987). "Tumors of the bone," pages 169 to 175 in *Pathology of Tumors of Laboratory Animals, Volume 1, Tumors of the Rat*, Turnsor, V.S., Ed. (International Agency for Research on Cancer, Lyon).

LLOYD, E.L., LOUTIT, J.F. and MACKEVICIUS, F. (1975). "Viruses in osteosarcomas induced by ^{226}Ra. A study of the induction of bone tumours in mice," Int. J. Radiat. Biol. Relat. Stud. Phys. Chem. Med. **28**, 13–33.

LLOYD, R.D., BRUENGER, F.W., MILLER, S.C., ANGUS, W., TAYLOR, G.N., JEE, W.S. and POLIG, E. (1991). "Distribution of radium-induced bone cancers in beagles and comparison with humans," Health Phys. **60**, 435–438.

LLOYD, R.D., TAYLOR, G.N., ANGUS, W., BRUENGER, F.W. and MILLER, S.C. (1993). "Bone cancer occurrence among beagles given^{239}Pu as young adults," Health Phys. **64**, 45–51.

LLOYD, R.D., MILLER, S.C., TAYLOR, G.N., BRUENGER, F.W., JEE, W.S. and ANGUS, W. (1994). "Relative effectiveness of ^{239}Pu and some other internal emitters for bone cancer induction in beagles," Health Phys. **67**, 346–353.

LLOYD, R.D., TAYLOR, G.N., MILLER, S.C., BRUENGER, F.W. and JEE, W.S.S. (2001). "Review of ^{239}Pu and ^{226}Ra effects in beagles," Health Phys. **81**, 691–697.

LO, Y.M., DARBY, S., NOAKES, L., WHITLEY, E., SILCOCKS, P.B.S., FLEMING, K.A. and BELL, J.I. (1995). "Letter: Screening for codon 249 *p53* mutation in lung cancer associated with domestic radon exposure," Lancet **345**, 60.

LOCK, L.F. and WICKRAMASINGHE, D. (1994). "Cycling with CDKs," Trends Cell Biol. **4**, 404–405.

LOEB, J. and NORTHROP, J.H. (1916). "Is there a temperature coefficient for the duration of life?" Proc. Natl. Acad. Sci. **2**, 456–457.

LOEB, J. and NORTHROP, J.H. (1917). "What determines the duration of life in metazoa?" Proc. Natl. Acad. Sci. **3**, 382–386.

LOPEZ, B. and COPPEY, J. (1987). "Promotion of double-strand break repair by human nuclear extracts preferentially involves recombination with intact homologous DNA," Nucleic Acids Res. **15**, 6813–6826.

LOPEZ, B.S., CORTEGGIANI, E., BERTRAND-MERCAT, P. and COPPEY, J. (1992). "Directional recombination is initiated at a double strand break in human nuclear extracts," Nucleic Acids Res. **20**, 501–506.

LORENZ, E. (1950). "Some biologic effects of long continued irradiation," Am. J. Roentgenol. Radium Ther. Nucl. Med. **63**, 176–185.

LORENZ, E., HESTON, W.E., ESCHENBRENNER, E.B. and DERINGER, M.K. (1947). "Biological studies in the tolerance range," Radiology **49**, 274–285.

LORENZ, E., JACOBSON, L.O., HESTON, W.E., SHIMKIN, M., ESCHENBRENNER, E.B., DERINGER, M.K., DONIGER, J. and SCHWEISTHAL, R. (1954). "Effects of long-continued total-body gamma irradiation on mice, guinea pigs and rabbits. III. Effects on life span, weight, blood picture and carcinogenesis and the role of the intensity of radiation" pages 24 to 148 in *Biological Effects of External X and Gamma Radiation, Vol. 1, National Nuclear Energy Series IV-22B*, R.E. Zirkle, Ed. (McGraw-Hill, New York).

LOWE, S.W., SCHMITT, E.M., SMITH, S.W., OSBORNE, B.A. and JACKS, T. (1993). "*p53* is required for radiation-induced apoptosis in mouse thymocytes," Nature **362**, 847–849.

LUBIN, J.H., BOICE, JR., J.D., EDLING, C., HORNUNG, R., HOWE, G., KUNZ, E., KUSIAK, R.A., MORRISON, H.I., RADFORD, E.P., SAMET, J.M., TIRMARCHE, M., WOODWARD, A., XIANG, Y.S. and PIERCE, D.A. (1994). *Radon and Lung Cancer Risk: A Joint Analysis of 11 Underground Miners Studies*, NIH Publication No. 94-3644 (U.S. Department of Health and Human Services, Washington).

LUBIN, J.H., BOICE, JR., J.D., EDLING, C., HORNUNG, R.W., HOWE, G.R., KUNZ, E., KUSIAK, R.A., MORRISON, H.I., RADFORD, E.P., SAMET, J.M., TIRMARCHE, M., WOODWARD, A., YAO, S.X. and PIERCE, D.A. (1995). "Lung cancer in radon-exposed miners and estimation of risk from indoor exposure," J. Natl. Cancer Inst. **87**, 817–826.

LUCAS, J.N., AWA, A., STRAUME, T., POGGENSEE, M., KODAMA, Y., NAKANO, M., OHTAKI, K., WEIER, H.U., PINKEL, D., GRAY, J. and LITTLEFIELD, L.G. (1992). "Rapid translocation frequency analysis in humans decades after exposure to ionizing radiation," Int. J. Radiat. Biol. **62**, 53–63.

LUCAS, J.N., HILL, F., BURK, C., FESTER, T. and STRAUME, T. (1995). "Dose-response curve for chromosome translocations measured in human lymphocytes exposed to ^{60}Co gamma rays," Health Phys. **68**, 761–765.

LUCKE-HUHLE, C., COMPER, W., HIEBER, L. and PECH, M. (1982). "Comparative study of G2 delay and survival after 241 Americium-alpha and 60 cobalt-gamma irradiation," Radiat. Environ. Biophys. **20**, 171–185.

LUDWIG, F.C., ELASHOFF, R.M. and WELLINGTON, J.S. (1968). "Murine radiation leukemia and the preleukemic state," Lab. Invest. **19**, 240–250.

LUNDGREN, D.L., HAHN, F.F., GRIFFITH, W.C. and BOECKER, B.B. (1992a). "Effects of thoracic and whole-body exposure of F344 rats to x rays," pages 121 to 122 in *Inhalation Toxicology Research Institute Annual Report 1991–1992*, Report 138 (Lovelace Biomedical and Environmental Research Institute, Albuquerque, New Mexico).

LUNDGREN, D.L., HAHN, F.F., GRIFFITH, W.C., CARLTON, W.W., HOOVER, M.D. and BOECKER, B.B. (1992b). "Effects of combined exposure of F344 rats to ^{239}PuO$_2$ and whole-body x-radiation," pages 115 to 116 in *Inhalation Toxicology Research Institute Annual Report 1991–1992*, Report 138, Lovelace Biomedial and Environmental Research Institute, Albuquerque, New Mexico.

LUNDGREN, D.L., HAHN, F.F. and DIEL, J.H. (1992c). "Repeated inhalation exposure of rats to aerosols of ^{144}CeO$_2$, II. Effects on survival and lung, liver, and skeletal neoplasms," Radiat. Res. **132**, 325–333.

LUNDGREN, D.L., HALEY, P.J., HAHN, F.F., DIEL, J.H., GRIFFITH, W.C. and SCOTT, B.R. (1995). "Pulmonary carcinogenicity of repeated inhalation exposure of rats to aerosols of $^{239}PuO_2$," Radiat. Res. **142**, 39–53.

LUNDGREN, D.L., HAHN, F.F., GRIFFITH, W.C., HUBBS, A.F., NIKULA, K.J., NEWTON, G.J., CUDDIHY, R.G. and BOECKER, B.B. (1996). "Pulmonary carcinogenicity of relatively low doses of beta-particle radiation from inhaled $^{144}CeO_2$ in rats," Radiat. Res. **146**, 525–535.

LUST, J.A. (1996). "Molecular genetics and lymphoproliferative disorders," J. Clin. Lab. Anal. **10**, 359–367.

MACHINAMI, R., ISHIKAWA, Y., and BOECKER, B.B., Eds. (1999). "The international workshop on the health effects of thorotrast, radium, radon and other alpha-emitters 1999," Radiat. Res. **152** (Suppl. 6).

MAKEHAM, W.H. (1867). "On the law of mortality," J. Inst. Actuaries **13**, 325–358.

MAKINO, H., ISHIZAKA, Y., TSUJIMOTO, A., NAKAMURA, T., ONDA, M., SUGIMURA, T. and NAGAO, M. (1992). "Rat *p53* gene mutations in primary Zymbal gland tumors induced by 2-amino-3-methylimidazo[4,5-f]quinoline, a food mutagen," Proc. Natl. Acad. Sci. USA **89**, 4850–4854.

MALAKOFF, D. (1999). "Bayes offers a 'new' way to make sense of numbers," Science **286**, 1460–1464.

MALKINSON, A.M. (1992). "Primary lung tumors in mice: An experimentally manipulable model of human adenocarcinoma," Cancer Res. **52**, 2670S–2676S.

MANCUSO, M., PAZZAGLIA, S., TANORI, M., HAHN, H., MEROLA, P., REBESSI, S., ATKINSON, M.J., DI MAJO, V., COVELLI, V. and SARAN, A. (2004). "Basal cell carcinoma and its development: Insights from radiation-induced tumors in *PTCH*1-deficient mice," Cancer Res. **64**, 934–941.

MANECKJEE, R. and MINNA, J.D. (1990). "Opioid and nicotine receptors affect growth regulation of human lung cancer cell lines," Proc. Natl. Acad. Sci. USA **87**, 3294–3298.

MANENTI, G., BINELLI, G., GARIBOLDI, M., CANZIAN, F., DE GREGORIO, L., FALVELLA, F.S., DRAGANI, T.A. and PIEROTTI, M.A. (1994). "Multiple loci affect genetic predisposition to hepatocarcinogenesis in mice," Genomics **23**, 118–124.

MARANGOS, P.J., GAZDAR, A.F. and CARNEY, D.N. (1982). "Neuron specific enolase in human small cell carcinoma cultures," Cancer Lett. **15**, 67–71.

MARKS, S. and SULLIVAN, M.F. (1960). "Tumours of the small intestine in rats after intestinal x-irradiation," Nature **188**, 953.

MARTLAND, H.S. (1929). "Occupational poisoning in the manufacture of luminous watch dials," J. Am. Med. Assoc. **92**, 466–473.

MAYS, C.W., DOUGHERTY, T.F., TAYLOR, G.N., LLOYD, R.D., STOVER, B.J., JEE, W.S.S., CHRISTENSEN, W.R., DOUGHERTY, J.H. and ATHERTON, D.R. (1969). "Radiation-induced bone cancer in beagles," page 393 in *Delayed Effects of Bone-Seeking Radionuclides,*" Mays,

C.W., Jee, W.S.S., Lloyd, R.D., Stover, B.J., Dougherty, J.H. and Taylor, G.N., Eds. (University of Utah Press, Salt Lake City, Utah).

MAYS, C.W., TAYLOR, G.N. and LLOYD, R.D. (1986a). "Toxicity ratios: Their use and abuse in predicting the risk from induced cancer in life-span radiation effects in animals. What can they tell us?" Thompson, R.C. and Mahaffey, J.A., Eds. (Office of Scientific and Technical Information, U.S. Department of Education, Springfield, Virginia).

MAYS, C.W., SPIESS, H., CHMELEVSKY, D. and KELLERER, A. (1986b). "Bone sarcoma cumulative tumor rates in patients injected with ^{224}Ra," pages 27 to 31 in *The Radiobiology of Radium and Thorotrast*, Gossner, W., Gerber, G.B., Hagen, U. and Luz, A., Eds. (Urban and Schwarzenberg, Baltimore).

MAYS, C.W., TAYLOR, G.N., LLOYD, R.D., BRUENGER, F.W. and ANGUS, W. (1989). "Bone sarcoma induction by radium-224 in beagles: An interim report," pages 47 to 54 in *Risks from Radium and Thorotrast*, Taylor, D.M., Mays, C.W., Gerber, G.B. and Thomas, R.G., Eds., BIR Report 21 (British Institute of Radiology, London).

MCDOWELL, E.M. (1987). "Bronchogenic carcinomas," page 255 in *Lung Carcinomas*, McDowell, E.M., Ed. (Elsevier Science, New York).

MCKENNA, W.G., WEISS, M.C., BAKANAUSKAS, V.J., SANDLER, H., KELSTEN, M.L., BIAGLOW, J., TUTTLE, S.W., ENDLICH, B., LING, C.C. and MUSCHEL, R.J. (1990). "The role of the *H-ras* oncogene in radiation resistance and metastasis," Int. J. Radiat. Oncol. Biol. Phys. **18**, 849–859.

MCMORROW, L.E., NEWCOMB, E.W. and PELLICER, A. (1988). "Identification of a specific marker chromosome early in tumor development in gamma-irradiated C57BL/6J mice," Leukemia **2**, 115–119.

MEDAWAR, P.B. (1952). *An Unsolved Problem of Biology* (H.K. Lewis, London).

MEDINA, D. and SMITH G.H. (1999). "Chemical carcinogen-induced tumorigenesis in parous, involuted mouse mammary gland," J. Natl. Cancer Inst. **91**, 967–969.

MEDINA, D. and WARNER, M.R. (1976). "Mammary tumorigenesis in chemical carcinogen-treated mice. IV. Induction of mammary ductal hyperplasias," J. Natl. Cancer Inst. **57**, 331–337.

MEIJNE, E., HUISKAMP, R., HAINES, J., MOODY, J., FINNON, R., WILDING, J., SPANJER, S., BOUFFLER, S., EDWARDS, A., COX, R. and SILVER, A. (2001). "Analysis of loss of heterozygosity in lymphoma and leukemia arising in F_1 hybrid mice locates a common region of chromosome 4 loss," Genes Chromosomes Cancer **31**, 373–381.

MELAMED, M.R. and ZAMAN, M.B. (1982). "Pathogenesis of epidermoid carcinoma of the lung," page 37 in *Morphogenesis of Lung Cancer*, Shimosato, Y., Melamed, M.R. and Nettesheim, P., Eds. (CRC Press, Boca Raton, Florida).

MELENDEZ, B., SANTOS, J. and FERNANDEZ-PIQUERAS, J. (1999). "Loss of heterozygosity at the proximal-mid part of mouse chromo-

some 4 defines two novel tumor suppressor gene loci in T-cell lymphomas," Oncolgene **18**, 4166–4169.
METZGER, L. and ILIAKIS, G. (1991). "Kinetics of DNA double-strand break repair throughout the cell cycle as assayed by pulsed field gel electrophoresis in CHO cells," Int. J. Radiat. Biol. **59**, 1325–1339.
MILDVAN, A. and STREHLER, B.L. (1960). "A critique of theories of mortality," pages 216 to 235 in *The Biology of Aging*, Strehler, B.L., Ebert, J.D., Glass, H.B. and Shock, N.W., Eds. (American Institute of Biological Sciences, Washington).
MILLER, C.W., ASLO, A., CAMPBELL, M.J., KAWAMATA, N., LAMPKIN, B.C. and KOEFFLER, H.P. (1996). "Alterations of the *p15*, *p16*, and *p18* genes in osteosarcoma," Cancer Genet. Cytogenet. **86**, 136–142.
MILLER, S.C., LLOYD, R.D., BRUENGER, F.W., KRAHENBUHL, M.P., POLIG, E. and ROMANOV, S.A. (2003). "Comparisons of the skeletal locations of putative plutonium-induced osteosarcomas in humans with those in beagle dogs and with naturally occurring tumors in both species," Radiat. Res. **160**, 517–523.
MINNA, J.D., PASS, H., GLATSTEIN, E. and IHDE, D.C. (1989). "Cancer of the lung", pages 591 to 705 in *Cancer: Principles and Practice of Oncology*, 3rd ed., DeVita, V.T., Jr., Hellman, S. and Rosenberg, S.A., Eds. (J.B. Lippincott Co., Philadelphia).
MITSUDOMI T., VIALLET, J., MULSHINE, J.L., LINNOILA, R.I., MINNA, J.D. and GAZDAR, A.F. (1991). "Mutations of *ras* genes distinguish a subset of non-small-cell lung cancer cell lines from small-cell lung cancer cell lines," Oncogene **6**, 1353–1362.
MOLE, R.H. (1955). "On wasted radiation and the interpretation of experiments with chronic irradiation," J. Natl. Cancer Inst. **15**, 907–914.
MOLE, R.H. (1984). "Dose-response relationships, pages 403 to 420 in *Radiation Carcinogenesis: Epidemiology and Biological Significance*, Boice, J.D., Jr. and Fraumeni, J.F., Jr., Eds. (Raven Press, New York).
MOLE, R.H., PAPWORTH, D.G. and CORP, M.J. (1983). "The dose-response for x-ray induction of myeloid leukaemia in male CBA/H mice," Br. J. Cancer **47**, 285–291.
MONTAGNA, W. (1962). *The Structure and Function of Skin* (Academia Press, New York).
MOODY, T.W., LEE, M., KRIS, R.M., BELLOT, F., BEPLER, G., OIE, H. and GAZDAR, A.F. (1990). "Lung carcinoid cell lines have bombesin-like peptides and EGF receptors," J. Cell Biochem. **43**, 139–147.
MOOLGAVKAR, S.H. and VENZON, D.J. (1979). "Two-event models for carcinogenesis: Incidence curves for childhood and adult tumors," Math Biosci. **47**, 55–77.
MOOLGAVKAR, S.H., DAY, N.E. and STEVENS, R.G. (1980). "Two-stage model for carcinogenesis: Epidemiology of breast cancer in females," J. Nat. Cancer Inst. **65**, 559–569.
MORRISON, P.F. (1987). "Effects of time-variant exposure on toxic substance response," Environ. Health Perspect. **76**, 133–139.

MOSER, A.R., PITOT, H.C. and DOVE, W.F. (1990). "A dominant mutation that predisposes to multiple intestinal neoplasia in the mouse," Science **247**, 322–324.

MOSER, A.R., LUONGO, C., GOULD, K.A., MCNELEY, M.K., SHOEMAKER, A.R. and DOVE, W.F. (1995). "ApcMin: A mouse model for intestinal and mammary tumorigenesis," Eur. J. Cancer **31A**, 1061–1064.

MOTHERSILL, C. and SEYMOUR, C. (2001). "Radiation-induced bystander effects: Past history and future directions," Radiat. Res. **155**, 759–767.

MRC (1956). Medical Research Council. *The Hazards to Man of Nuclear and Allied Radiations: A Report to the Medical Research Council* (Her Majesty's Stationary Office, London).

MUGGENBURG, B.A., MEWHINNEY, J.A., GRIFFITH, W.C., HAHN, F.F., MCCLELLAN, R.O., BOECKER, B.B. and SCOTT, B.R. (1983). "Dose-response relationships for bone cancers from plutonium in dogs and people," Health Phys. **44** (Suppl. 1), 529–535.

MUGGENBURG, B.A., GUILMETTE, R.A., MEWHINNEY, J.A., GILLETT, N.A., MAUDERLY, J.L., GRIFFITH, W.C., DIEL, J.H., SCOTT, B.R., HAHN, F.F. and BOECKER, B.B. (1996a). "Toxicity of inhaled plutonium dioxide in beagle dogs," Radiat. Res. **145**, 361–381.

MUGGENBURG, B.A., HAHN, F.F., GRIFFITH, W.C., JR., LLOYD, R.D. and BOECKER, B.B. (1996b). "The biological effects of radium-224 injected into dogs," Radiat. Res. **146**, 171–186.

MULCAHY, R.T., GOULD, M.N. and CLIFTON, K.H. (1984). "Radiogenic initiation of thyroid cancer: A common cellular event," Int. J. Radiat. Biol. Relat. Stud. Phys. Chem. Med. **45**, 419–426.

MULLER, H.J. (1927). "Artificial transmutation of the gene," Science **66**, 84–87.

MULLER, H.J. (1950). "Our load of mutations," Am. J. Hum. Genet. **2**, 111–176.

MUNGER, K. (2002). "Disruption of oncogene/tumor suppressor networks during human carcinogenesis," Cancer Invest. **20**, 71–81.

MURRAY, A.W. and HUNT, T. (1993). *The Cell Cycle: An Introduction* (W.H. Freeman and Co., New York).

MUTO, M., KUBO, E., SADO, T. and YAMAGISHI, H. (1991). "Characterization of thymic prelymphoma cells that develop during radiation-induced lyphomagenesis in B10 mice." J. Radiat. Res. (Tokyo) **32** (Suppl. 2), 156–167.

NAKAMURA, N. and OKADA, S. (1981). "Dose-rate effects of gamma-ray-induced mutations in cultured mammalian cells," Mutat. Res. **83**, 127–135.

NAS/NRC (1956). National Academy of Sciences/National Research Council. *The Biological Effects of Atomic Radiation, Summary Reports from a Study* (National Academy Press, Washington).

NAS/NRC (1972). National Academy of Sciences/National Research Council. *The Effects on Populations of Exposure to Low Levels of Ionizing*

Radiation: BEIR I, Committee on the Biological Effects of Ionizing Radiations (National Academy Press, Washington).

NAS/NRC (1975). National Academy of Science/National Research Council. *Pest Control, an Assessment of Present and Alternative Technologies: Study on Problems of Pest Control. Vol. 1, Contemporary Pest Control Practices and Prospects* (National Academy Press, Washington).

NAS/NRC (1980). National Academy of Sciences/National Research Council. *The Effect of Exposure to Low Levels of Ionizing Radiation: BEIR III*, Committee on the Biological Effects of Ionizing Radiations (National Academy Press, Washington).

NAS/NRC (1987). National Academy of Science/National Research Council. "Pharmacokinetics in risk assessment," page 512 in *Drinking Water and Health*, Vol. 8 (National Academy Press, Washington).

NAS/NRC (1988). National Academy of Sciences/National Research Council. *Health Risks of Radon and Other Internally Deposited Alpha-Emitters: BEIR IV*, Committee on the Biological Effects of Ionizing Radiations (National Academy Press, Washington).

NAS/NRC (1990). National Academy of Sciences/National Research Council. "Genetic effects of radiation," pages 65 to 134 in *Health Effects of Exposure to Low Levels of Ionizing Radiation: BEIR V*, Committee on the Biological Effects of Ionizing Radiations (National Academy Press, Washington).

NAS/NRC (1991). National Academy of Sciences/National Research Council. *Comparative Dosimetry of Radon in Mines and Homes* (National Academy Press, Washington).

NAS/NRC (1999). National Academy of Sciences/National Research Council. *Health Effects of Exposure to Radon: BEIR VI*, Committee on the Biological Effects of Ionizing Radiations (National Academy Press, Washington).

NATARAJAN, A.T. and OBE, G. (1978). "Molecular mechanisms involved in the production of chromosomal aberrations. I. Utilization of *Neurospora* endonuclease for the study of aberration production in G2 stage of the cell cycle," Mutat. Res. **52**, 137–149.

NATHRATH, M.H., KUOSAITE, V., ROSEMANN, M., KREMER, M. POREMBA, C., WAKANA, S., YANAGI, M., NATHRATH, W.B., HOFLER, H., IMAI, K. and ATKINSON, M.J. (2002). "Two novel tumor suppressor gene loci on chromosome 6q and 15q in human osteosarcoma identified through comparative study of allelic imbalances in mouse and man," Oncogene **21**, 5975–5980.

NAU, M.M., BROOKS, B.J., JR., CARNEY, D.N., GAZDAR, A.F., BATTEY, J.F., SAUSVILLE, E.A. and MINNA, J.D. (1986). "Human small-cell lung cancers show amplification and expression of the N-*myc* gene," Proc. Natl. Acad. Sci. USA **83**, 1092–1096.

NCBI (2005). National Center for Biotechnology Information. *Genes and Disease* (online), http://www.ncbi.nlm.nih.gov/books/bv.fcgi?call=bv. View.ShowSection&rid=gnd.preface.91 (accessed September 2005)

(National Center for Biotechnology Information, National Library of Medicine, National Institutes of Health, Bethesda, Maryland).
NCRP (1980). National Council on Radiation Protection and Measurements. *Influence of Dose and Its Distribution in Time on Dose-Response Relationships for Low-LET Radiations*, NCRP Report No. 64 (National Council on Radiation Protection and Measurements, Bethesda, Maryland).
NCRP (1991). National Council on Radiation Protection and Measurements. *Some Aspects of Strontium Radiobiology*, NCRP Report No. 110 (National Council on Radiation Protection and Measurements, Bethesda, Maryland).
NCRP (1993). National Council on Radiation Protection and Measurements. *Limitation of Exposure to Ionizing Radiation*, NCRP Report No. 116 (National Council of Radiation Protection and Measurements, Bethesda, Maryland).
NCRP (1997). National Council on Radiation Protection and Measurements. *Uncertainties in Fatal Cancer Risk Estimates Used in Radiation Protection*, NCRP Report No. 126 (National Council of Radiation Protection and Measurements, Bethesda, Maryland).
NCRP (1998). National Council on Radiation Protection and Measurements. *Radionuclide Exposure of the Embryo/Fetus*, NCRP Report No. 128 (National Council on Radiation Protection and Measurements, Bethesda, Maryland).
NCRP (2001a). National Council on Radiation Protection and Measurements. *Liver Cancer Risk from Internally-Deposited Radionuclides*, NCRP Report No. 135 (National Council on Radiation Protection and Measurements, Bethesda, Maryland).
NCRP (2001b) National Council on Radiation Protection and Measurements. *Evaluation of the Linear-Nonthreshold Dose-Response Model for Ionizing Radiation*, NCRP Report No. 136 (National Council on Radiation Protection and Measurements, Bethesda, Maryland).
NEEL, J.V., SCHULL, W.J., AWA, A.A., SATOH, C., KATO, H., OTAKE, M. and YOSHIMOTO, Y. (1990). "The children of parents exposed to atomic bombs: Estimates of the genetic doubling dose of radiation for humans," Am. J. Hum. Genet. **46**, 1053–1072.
NEJM (1992). New England Journal of Medicine. "Correspondence," N. Engl. J. Med. **326**, 1357–1361.
NEKOLLA, E.A., KELLERER, A.M., KUSE-ISINGSCHULTE, M., EDER, E. and SPIESS, H. (1999). "Malignancies in patients treated with high doses of radium-224," Radiat. Res. **152** (Suppl. 6), S3–S7.
NIKULA, K.J., MUGGENBURG, B.A., CHANG, I.Y., GRIFFITH, W.C., HAHN, F.F. and BOECKER, B.B. (1995). "Biological effects of ^{137}CsCl injected in beagle dogs," Radiat. Res. **142**, 347–361.
NILSSON, A. and STANTON, M.F. (1994). "Tumors of the bone," pages 681 to 729 in *Pathology of Tumours in Laboratory Animals, Volume 2, Tumours of the Mouse,* 2nd ed., IARC Scientific Publication No. 111, Turosov, V.S. and Mohr, U., Eds. (International Agency for Research on Cancer, Lyon).

NIWA, O., ENOKI, Y. and YOKORO, K. (1989). "Overexpression and amplification of the c-*myc* gene in mouse tumors induced by chemicals and radiations," Jpn. J. Cancer Res. **80**, 212–218.

NORRIS, W.P., TYLER, S.A. and SACHER, G.A. (1976). "An interspecies comparison of responses of mice and dogs to continuous ^{60}Co-irradiation," pages 147 to 156 in *Biological and Environmental Effects of Low-Level Radiation*, STI/PUB/409 (International Atomic Energy Agency, Vienna).

NOWELL, P.C. and COLE, L.J. (1959). "Late effects of fast neutrons versus x-rays in mice: Nephroselerosis, tumors, longevity," Radiat. Res. **11**, 545–556.

NOWELL, P.C., COLE, L.G. and ELLIS, M.F. (1956). "Induction of intestinal carcinoma in the mouse by whole-body fast-neutron irradiation," Cancer Res. **16**, 873–876.

NYLEN, E.S., BECKER, K.L., JOSHI, P.A., SNIDER, R.H. and SCHULLER, H.M. (1990). "Pulmonary bombesin and calcitonin in hamsters during exposure to hyperoxia and diethylnitrosamine," Am. J. Respir. Cell Mol. Biol. **2**, 25–31.

OHNO, S., MIGITA, S., WIENER, F., BABONITS, M., KLEIN, G., MUSHINSKI, J.F. and POTTER, M. (1984). "Chromosomal translocations activating *myc* sequences and transduction of v-*abl* are critical events in the rapid induction of plasmacytomas by pristane and abelson virus," J. Exp. Med. **159**, 1762–1777.

OLSHANSKY, S.J. and CARNES, B.A. (1997). "Ever since Gompertz," Demography **34**, 1–15.

OLSHANSKY, S.J., CARNES, B.A. and CASSEL, C.K. (1993). "Letter: Fruit fly aging and mortality," Science **260**, 1565–1567.

OREFFO, V.I.C., LIN, H.W., PADMANABHAN, R. and WITSCHI, H. (1993). "K-*ras* and *p53* point mutations in 4-(methylnitrosamino)-1-(3-pyridyl)-1-butanone-induced hamster lung tumors," Carcinogenesis **14**, 451–455.

OSBORNE, J.W., NICHOLSON, D.P. and PRASAD, K.N. (1963). "Induction of intestinal carcinoma in the rat by x-irradiation of the small intestine," Radiat. Res. **18**, 76–85.

PAGE, D.L. and ANDERSON, T.J., Eds. (1987). *Diagnostic Histopathology of the Breast* (Elsevier Science, New York).

PARDO, F.S., SU, M., BOREK, C., PREFFER, F., DOMBKOWSKI, D., GERWECK, L. and SCHMIDT, E.V. (1994). "Transfection of rat embryo cells with mutant *p53* increases the intrinsic radiation resistance," Radiat. Res. **140**, 180–185.

PARK, J.F. and STAFF (1993). *Pacific Northwest Laboratory Annual Report for 1992 to the DOE Office of Energy Research, Part 1: Biomedical Sciences*, Report PNL-8500, Pt. 1 (Pacific Northwest National Laboratory, Richland, Washington).

PARK, J.F., BUSCHBOM, R.L., DAGLE, G.E., JAMES, A.C., WATSON, C.R. and WELLER, R.E. (1997). "Biological effects of inhaled ^{238}PuO$_2$ in beagles," Radiat. Res. **148**, 365–381.

PARSHAD, R., SANFORD, K.K. and JONES, G.M. (1983). "Chromatid damage after G2 phase x-irradiation of cells from cancer-prone individuals implicates deficiency in DNA repair," Proc. Natl. Acad. Sci. USA **80**, 5612–5616.

PAWEL, D.J. and PUSKIN, J.S. (2004). "The U.S. Environmental Protection Agency's assessment of risks from indoor radon," Health Phys. **87**, 68–74.

PEARL, R.A. (1922). "Comparison of the laws of mortality in *Drosophila* and man," Am. Nat. **56**, 398–405.

PEARL, R.A. (1923). "A comparison of the mortality of certain lower organisms with that of man," Science **57**, 209–212.

PEARL, R.A. and MINER, J.R. (1935). "Experimental studies on the duration of life: XIV. The comparative mortality of certain lower organisms," Quart. Rev. Biol. **10**, 60–79.

PEGG, A.E. (2000). "Repair of 0(6)-alkylguanine by alkyltransferases," Mutat. Res. **462**, 83–100.

PELICCI, P.G., KNOWLES, D.M., II, MAGRATH, I. and DALLA-FAVERA, R. (1986). "Chromosomal breakpoints and structural alterations of the c-*myc* locus differ in endemic and sporadic forms of Burkitt lymphoma," Proc. Natl. Acad. Sci. USA **83**, 2984–2988.

PERKINS, A.C., and CORY, S. (1993). "Conditional immortalization of mouse myelomonocytic, megakaryocytic and mast cell progenitors by the *Hox-2.4* homeobox gene," EMBO J. **12**, 3835–3846.

PERKINS, A., KONGSUWAN, K., VISVADER, J., ADAMS, J.M. and CORY, S. (1990). "Homeobox gene expression plus autocrine growth factor production elicits myeloid leukemia," Proc. Natl. Acad. Sci. USA **87**, 8398–8402.

PETO, R., PIKE, M.C., BERNSTEIN, L., GOLD, L.S. and AMES, B.N. (1984). "The TD_{50}: A proposed general convention for the numerical description of the carcinogenic potency of chemicals in chronic-exposure animal experiments," Environ. Health Perspect. **58**, 1–8.

PFEIFFER, P. and VIELMETTER, W. (1988). "Joining of nonhomologous DNA double strand breaks *in vitro*," Nucleic Acids Res. **16**, 907–924.

PIERCE, D.A., SHIMIZU, Y., PRESTON, D.L., VAETH, M. and MABUCHI, K. (1996). "Studies of the mortality of atomic bomb survivors. Report 12, Part I. Cancer: 1950–1990," Radiat. Res. **146**, 1–27.

PINKEL, D. (1958). "The use of body surface area as a criterion of drug dosage in cancer chemotherapy," Cancer Res. **18**, 853–856.

PINTO, M., PRISE, K.M. and MICHAEL, B.D. (2002). "Double strand break rejoining after irradiation of human fibroblasts and x rays or alpha particles: PFGE studies and numerical models," Radiat. Prot. Dosim. **99**, 133–136.

PLOPPER, C.G., HILL, L.H. and MARIASSY, A.T. (1980). "Ultrastructure of the nonciliated bronchiolar epithethial (Clara) cell of mammalian lung: III, A study of man with comparison of 15 mammalian species," Exp. Lung Res. **1**, 171–180.

POPLACK, D.G., KUN, L.E., CASSADY, J.R. and PIZZO, P.A. (1989). "Leukemias and lymphomas of childhood," pages 1671 to 1695 in

Cancer: Principles and Practice of Oncology, DeVita, V.T., Jr., Hellman, S. and Rosenberg, S.A., Eds. (J.B. Lippincolt Company, Philadelphia).

PRELEC, D. (2004). "A Bayesian truth serum for subjective data," Science **306**, 462–466.

PRESTON, R.J. (1992). "Commentary to Thacker: A consideration of the mechanisms of induction of mutations in mammalian cells by low doses and dose rates of ionizing radiation," pages 125 to 135 in *Advances in Radiation Biology. Effects of Low Dose and Low Dose Rate Radiation, Volume 16*, Nygaard, O.F., Sinclair, W.K. and Lett, J.T., Eds. (Academic Press, Inc., New York).

PRESTON, D.L., KUSUMI, S., TOMONAGA, M., IZUMI, S., RON, E., KURAMOTO, A., KAMADA, N., DOHY, H., MATSUO, T., MATSUI, T., THOMPSON, D.E., SODA, M. and MABUCHI, K. (1994). "Cancer incidence in atomic bomb survivors. Part III. Leukemia, lymphoma and multiple myeloma," Radiat. Res. **137**, S68–S97.

PRESTON, D.L., MATTSSON, A., HOLMBERG, E., SHORE, R., HILDRETH, N.G. and BOICE, J.D., JR. (2002). "Radiation effects on breast cancer risk: A pooled analysis of eight cohorts," Radiat. Res. **158**, 220–235.

PRESTON, D.L., SHIMIZU, Y., PIERCE, D.A., SUYAMA, A. and MABUCHI, K. (2003). "Studies of mortality of atomic bomb survivors, Report 13: Solid cancer and noncancer disease mortality: 1950–1997," Radiat. Res. **160**, 381–407.

RAABE, O.G. (1989). "Scaling of fatal cancer risks from laboratory animals to man," Health Phys. **57** (Suppl. 1), 419–432.

RAABE, O.G. (1994). "Three-dimensional models of risk from internally deposited radionuclides," pages 633 to 656 in *Internal Radiation Dosimetry*, Raabe, O.G., Ed. (Medical Physics Publishing, Madison, Wisconsin).

RAABE, O.G., BOOK, S.A. and PARKS, N.J. (1983). "Lifetime bone cancer dose-response relationships in beagles and people from skeletal burdens of ^{236}Ra and ^{90}Sr," Health Phys. **44** (Suppl. 1), 33–48.

RABBITTS, T.H. (1994). "Chromosomal translocations in human cancer," Nature **372**, 143–149.

RABES, H.M. and KLUGBAUER, S. (1998). "Molecular genetics of childhood papillary thyroid carcinomas after irradiation: High prevalence of *RET* rearrangement," Recent Results Cancer Res. **154**, 248–264.

REVELL, S.H. (1974). "The breakage-and-reunion theory and the exchange theory for chromosomal aberrations induced by ionizing radiations: A short history," pages 367 to 416 in *Advances in Radiation Biology, Volume 4*, Lett, J.T., Adler, H. and Zelle, M., Eds. (Academic Press, Inc., New York).

REZNIK-SCHULLER, H. and REZNIK, G. (1979). "Experimental pulmonary carcinogenesis," Int. Rev. Exp. Pathol. **20**, 211–281.

RIGHI, M., SASSANO, M., VALSASNINI, P., SHAMMAH, S. and RICCIARDI-CASTAGNOLI, P. (1991). "Activation of the M-CSF gene in mouse macrophages immortalized by retroviruses carrying a v-*myc* oncogene," Oncogene **6**, 103–111.

RODENHUIS, S., SLEBOS, R.J.C., BOOT, A.J.M., EVERS, S.G., MOOI, W.J., WAGENAAR, S.S., VAN BODEGOM, P.C. and BOX, J.L. (1988). "Incidence and possible clinical significance of K-*ras* oncolgene activation in adenocarcinoma of the human lung," Cancer Res. **48**, 5738–5741.

RON, E. (1996). "The epidemiology of thyroid cancer" pages 1000 to 1020 in *Cancer Epidemiology and Prevention*, 2nd ed., Schottenfeld, D. and Fraumeni, J.F., Jr., Eds. (Oxford University Press, New York).

RON, E., PRESTON, D.L., KISHIKAWA, M., KOBUKE, T., ISEKI, M., TOKUOKA, S., TOKUNAGO, M. and MABUCHI, K. (1998). "Skin tumor risk among atomic-bomb survivors in Japan," Cancer Causes Control **9**, 393–401.

ROOTS, R., HOLLEY, W., CHATTERJEE, A., IRIZARRY, M. and KRAFT, G. (1990). "The formation of strand breaks in DNA after high-LET irradiation: A comparison of data from *in vitro* and cellular systems," Int. J. Radiat. Biol. **58**, 55–69.

ROSELL, R., LI, S., SKACEL, Z., MATE, J.L., MAESTRE, J., CANELA, M., TOLOSA, E., ARMENGOL, P., BARNADAS, A. and ARIZA, A. (1993). "Prognostic impact of mutated K-*ras* gene in surgically resected non-small cell lung cancer patients," Oncogene **8**, 2407–2412.

ROSEMANN, M., KUOSAITE, V., NATHRATH, M. and ATKINSON, M.J. (2002). "The genetics of radiation-induced and sporadic osteosarcoma: A unifying theory," J. Radiol. Prot. **22**, A113–A116.

ROSENBERG, N. and WITTE, O.N. (1988). "The viral and cellular forms of the Abelson (*abl*) oncogene," Adv. Virus Res. **35**, 39–81.

ROTHKAMM, K. and LOBRICH, M. (2002). "Misrepair of radiation-induced DNA double-strand breaks and its relevance for tumorigenesis and cancer treatment (review)," Int. J. Oncol. **21**, 433–440.

ROTHKAMM, K. and LOBRICH, M. (2003). "Evidence for a lack of DNA double-strand break repair in human cells exposed to very low x-ray doses," Proc. Natl. Acad. Sci. USA **100**, 5057–5062.

ROTMAN, G. and SHILOH, Y. (1998). "ATM: From gene to function," Hum. Mol. Genet. **7**, 1555–1563.

ROWLAND, R.E. (1994). *Radium in Humans. A Review of U.S. Studies* (online), Argonne National Laboratory Report ANL/ER-3, http://www.ipd.anl.gov/anlpubs/1994/11/16311.pdf (accessed September 2005) (National Technical Information Service, Springfield, Virginia).

ROWLAND, R.E. (1995). "Dose-response relationships for female radium dial workers: A new look," pages 135 to 43 in *Health Effects of Internally Deposited Radionuclides: Emphasis on Radium and Thorium*, van Kaick, G., Karaoglou, A. and Kellerer, A.M., Eds. (World Scientific, Hackensack, New Jersey).

ROWLAND, R.E., STEHNEY, A.F. and LUCAS, H.F. (1983). "Dose-response relationships for radium-induced bone sarcomas," Health Phys. **44** (Suppl. 1), 15–31.

ROWLEY, J.D. (1985). "Chromosome abnormalities in human leukemia as indicators of mutagenic exposure," pages 409 to 431 in *Carcinogenesis — A Comprehensive Survey, Volume 10, The Role of Chemicals and*

Radiation in the Etiology of Cancer, Huberman, E. and Barr, S.H., Eds. (Raven Press, New York).

ROWLEY, J.D. and TESTA, J.R. (1982). "Chromosome abnormalities in malignant hematologic diseases," Adv. Cancer Res. **36**, 103–148.

RUIVENKAMP, C.A.L., CSIKOS, T., KLOUS, A.M., VAN WEZEL, T. and DEMANT, P. (2003). "Five new mouse susceptibility to colon cancer loci, *Scc11-Scc15*," Oncogene **22**, 7258–7260.

RUSSELL, L.B. and RUSSELL, W.L. (1996). "Spontaneous mutations recovered as mosaics in the mouse specific-locus test," Proc. Nat. Acad. Sci. USA **93**, 13072–13077.

RUSSO, J. and RUSSO, I.H. (1987). "Biological and molecular bases of mammary carcinogenesis," Lab. Invest. **57**, 112–137.

SACCOMANNO, G., AUERBACH, O., KUSCHNER, M., HARLEY, N.H., MICHELS, R.Y., ANDERSON, M.W. and BECHTEL, J.J. (1996). "A comparison between the localization of lung tumors in uranium miners and in nonminers from 1947–1991," Cancer **77**, 1278–1283.

SACHER, G.A. (1950a). *The Survival of Mice Under Duration-of-Life Exposure to X-Rays at Various Dose Rates*, CH-3900 (University of Chicago, Chicago).

SACHER, G.A. (1950b). "Preliminary report: A comparative study of radiation lethality in several species of experimental animals irradiated for the duration-of-life," pages 105 to 122 in *Division of Biological and Medical Research Quarterly Report,* ANL-4488 (Argonne National Laboratory, Argonne, Illinois).

SACHER, G.A. (1955). "A comparative analysis of radiation lethality in mammals exposed at constant average intensity for the duration of life," J. Natl. Cancer Inst. **15**, 1125–1144.

SACHER, G.A. (1956a). "Survival of mice under duration-of-life exposure to x-rays at various rates," pages 435 to 463 in *Biological Effects of External X and Gamma Radiation, Part 2*, Zirkle, R.E., Ed. (McGraw-Hill, New York).

SACHER, G.A. (1956b). "On the statistical nature of mortality, with especial reference of chronic radiation mortality," Radiology **67**, 250–258.

SACHER, G.A. (1960). "Problems in the extrapolation of long-term effects from animals to man," pages 3 to 10 in *Symposium on the Delayed Effects of Whole-Body Radiation*, Watson, B.B., Ed. (Johns Hopkins University Press, Baltimore, Maryland).

SACHER, G.A. (1966). "The Gompertz transformation in the study of the injury-mortality relationship: Application to late radiation effects and aging," pages 411 to 441 in *Radiation and Aging*, Lindop, P.J. and Sacher, G.A., Eds. (Taylor and Francis, New York).

SACHER, G.A. and GRAHN, D. (1964). "Survival of mice under duration-of-life exposure to gamma rays. I. The dosage-survival relation and the lethality function," J. Natl. Cancer Inst. **32**, 277–321.

SACHER, G.A. and TRUCCO, E. (1962). "The stochastic theory of mortality," Ann. NY Acad. Sci. **96**, 985–1007.

SADO, T., KAMISAKU, H. and KUBO, E. (1991). "Bone marrow-thymus interactions during thymic lymphomagenesis induced by fractionated radiation exposure in B10 mice: Analysis using bone marrow transplantation between Thy 1 congenic mice," J. Radiat. Res. **32** (Suppl. 2), 168–180.

SAMID, D., MILLER, A.C., RIMOLDI, D., GAFNER, J. and CLARK, E.P. (1991). "Increased radiation resistance in transformed and nontransformed cells with elevated *ras* proto-oncogene expression," Radiat. Res. **126**, 244–250.

SANCAR, A. (1994). "Mechanisms of DNA excision repair," Science **266**, 1954–1956.

SANCAR, A. (1995). "Excision repair in mammalian cells," J. Biol. Chem. **270**, 15915–15918.

SANDBERG, A.A. (1993). "Chromosome changes in leukemia and cancer and their molecular limning," pages 141 to 163 in *The Causes and Consequences of Chromosomal Aberrations*, Kirsch, I.R., Ed. (CRC Press, Boca Raton, Florida).

SANDERS, C.L. and LUNDGREN, D.L. (1995). "Pulmonary carcinogenesis in the F344 and Wistar rat after inhalation of plutonium dioxide," Radiat. Res. **144**, 206–214.

SANDERS, C.L., LAUHALA, K.E., MCDONALD, K.E. and SANDERS, G.A. (1993a). "Lifespan studies in rats exposed to $^{239}PuO_2$ aerosol," Health Phys. **64**, 509–521.

SANDERS, C.L., LAUHALA, K.E. and MCDONALD, K.E. (1993b). "Lifespan studies in rats exposed to $^{239}PuO_2$ Aerosol. III. Survival and lung tumours," Int. J. Radiat. Biol. **64**, 417–430.

SANFORD, K.K., PARSHAD, R., GANTT, R., TARONE, R.E., JONES, G.M. and PRICE, F.M. (1989). "Factors affecting and significance of G2 chromatin radiosensitivity in predisposition to cancer," Int. J. Radiat. Biol. **55**, 963–981.

SANKARANARAYANAN, K. (1991a). "Ionizing radiation and genetic risks. III. Nature of spontaneous and radiation-induced mutations in mammalian *in vitro* systems and mechanisms of induction of mutations by radiation," Mutat. Res. **258**, 75–97.

SANKARANARAYANAN, K. (1991b). "Ionizing radiation and genetic risks. IV. Current methods, estimates of risk of Mendelian disease, human data and lessons from biochemical and molecular studies of mutation," Mutat. Res. **258**, 99–122.

SATOH, M.S., JONES, C.J., WOOD, R.D. and LINDAHL, T. (1993). "DNA excision-repair defect of xeroderma pigmentosum prevents removal of a class of oxygen free radical-induced base lesions," Proc. Natl. Acad. Sci. USA **90**, 6335–6339.

SAVITSKY, K., BAR-SHIRA, A., GILAD, S., ROTMAN, G., ZIV, Y., VANAGAITE, L., TAGLE, D.A., SMITH, S., UZIEL, T., SFEZ, S., ASHJENAZI, M., PECKER, I., FRYDMAN, M., HARNIK, R., PATANJALI, S.R., SIMMONS, A., CLINES, G.A., SATEIL, A., GATTI, R.A., CHESSA, L., SANAL, O., LAVIN, M.F., JASPERS, N.G.J., TAYLOR, A.M.R., ARLETT, C.F., MIKI, T., WEISSMAN, S.M., LOVETT, M., COLLINS, F.S. and SHILOH, Y. (1995). "A single ataxia telangiectasia gene with a product similar to PI-3 kinase," Science **268**, 1749–1753.

SAWYERS, C.L. (1992). *"bcr-abl* gene in chronic myelogenous leukemia," pages 37 to 51 in *Oncogenes in the Development of Leukaemia. Advances and Prospects in Clinical, Epidemiological and Laboratory Oncology: Cancer Surveys 15,* Witte, O.N., Ed. (Cold Spring Harbor Laboratory Press, Woodbury, New York).

SAX, K. (1938). "Chromosome aberrations induced by x-rays," Genetics **23**, 494–516.

SCHAEFFER, L., MONCOLLIN, V., ROY, R., STAUB, A., MEZZINA, M., SARASIN, A., WEEDA, G., HOEIJMAKERS, J.H. and EGLY, J.M. (1994). "The ERCC2/DNA repair protein is associated with the class II BTF2/TFIIH transcription factor," EMBO J. **13**, 2388–2392.

SCHNEIDERMANN, M.M., MANTEL, N. and BROWN, C.C. (1975). "From mouse to man—Or how to get from the laboratory to Park Avenue and 59th Street," Ann. N.Y. Acad. Sci. **246**, 237–248.

SCHON, A., MICHIELS, L., JANOWSKI, M., MERREGAERT, J. and ERFLE, V. (1986). "Expression of protooncogenes in murine osteosarcomas," Int. J. Cancer **38**, 67–74.

SCHULLER, H.M. (1989). "Comparative ultrastructural pathology of lung tumors," pages 1 to 41 in *Comparative Ultrastructural Pathology of Selected Tumors in Man and Animals,* Schuller, H.M., Ed. (CRC Press, Boca Raton, Florida).

SCHULLER, H.M. (1991). "Receptor-mediated mitogenic signals and lung cancer," Cancer Cells **3**, 496–503.

SCHULLER, H.M., BECKER, K.L. and WITSCHI, H.P. (1988). "An animal model for neuroendocrine lung cancer," Carcinogenesis **9**, 293–296.

SCHULLER, H.M., CORREA, E., ORLOFF, M. and REZNIK, G.K. (1990). "Successful chemotherapy of experimental neuroendocrine lung tumors in hamsters with an antagonist of Ca^{2+}/calmodulin," Cancer Res. **50**, 1645–1649.

SCOTT, D., SPREADBOROUGH, A., LEVINE, E. and ROBERTS, S.A. (1994). "Letter: Genetic predisposition in breast cancer," Lancet **344**, 1444.

SCOTT, D., SPREADBOROUGH, A.R., JONES, L.A., ROBERTS, S.A. and MOORE, C.J. (1996). "Chromosomal radiosensitivity in G2-phase lymphocytes as an indicator of cancer predisposition," Radiat. Res. **145**, 3–16.

SCOTTO, J., FEARS, T.R. and FRAUMENI, J.F., JR. (1983). *Incidence of Nonmelanoma Skin Cancer in the United States* (online), http://www.ciesin.org/docs/001-526/001.-526.html (assessed September 2005) U.S. Department of Health and Human Services, NIH Publication No. 83-2433 (National Institutes of Health, Bethesda, Maryland).

SEED, T.M. (1991). "Hematopoietic cell crisis: An early stage of evolving myeloid leukemia following radiation exposure," J. Radiat. Res. (Tokyo) **32** (Suppl. 2), 118–131.

SEED, T.M., KASPAR, L.V., TOLLE, D.V., FRITZ, T.E. and FRAZIER, M.E. (1988). "Analyses of critical target cell responses during preclinical phases of evolving chronic radiation-induced myeloproliferative disease: Exploitation of unique canine model," pages 245 to 255 in *Multilevel Health Effects Research: From Molecules to Man, Proceedings of the 27th Hanford Symposium on Health the Environment*, Park, J.F. and Pelroy, R.A., Eds. (Battelle Press, Columbus, Ohio).

SEIFTER, E.J. and IHDE, D.C. (1988). "Small cell lung cancer. A distinct clinicopathological entity," in *Lung Cancer: A Comprehensive Treatise*, Bitran, J.D., Colomb, H.M., Little, A.G. and Weichselbaum, R.R., Eds. (Grune and Stratton, New York).

SELBY, P.B. (1998a). "Major impacts of gonadal mosaicism on hereditary risk estimation, origin of hereditary diseases, and evolution," Genetica **102–103**, 445–462.

SELBY, P.B. (1998b). "Discovery of numerous clusters of spontaneous mutations in the specific-locus test in mice necessitates major increases in estimates of doubling dose," Genetica **102–103**, 463–487.

SELBY, C.P. and SANCAR, A. (1993). "Molecular mechanism of transcription-repair coupling," Science **260**, 53–58.

SELBY, P.B. and SELBY, P.R. (1977). "Gamma-ray-induced dominant mutations that cause skeletal abnormalities in mice. I. Plan, summary of results and discussion," Mutat. Res. **43**, 357–375.

SETO, E., USHEVA, A., ZAMBETTI, G.P., MOMAND, J., HORIKOSHI, N., WEINMANN, R., LEVINE, A.J. and SHENK, T. (1992). "Wild-type *p53* binds to the TATA-binding protein and represses transcription," Proc. Natl. Acad. Sci. USA **89**, 12028–12032.

SHELLABARGER, C.J. (1976). "Modifying factors in rat mammary gland carcinogenesis," pages 31 to 43 in *Biology of Radiation Carcinogenesis*, Yuhas, J.M., Tenant, R.W. and Regan, J.D., Eds. (Raven Press, New York).

SHIFRINE, M., BULGIN, M.S., DOLLARHIDE, N.E., WOLF, H.G., TAYLOR, N.J., WILSON, F.D., DUNGWORTH, D.L. and ZEE, Y.C. (1971). "Transplantation of radiation-induced canine myelomonocytic leukaemia," Nature **232**, 405–406.

SHILNIKOVA, N.S., PRESTON, D.L., RON, E., GILBERT, E.S., VASSILENKOV, E.K., ROMANOV, S.A., KUZNETSOVA, I.S., SOKOLNIKOV, M.E., OKATENKO, P.V., KRESLOV, V.V. and KOSHURNIKOVA, N.A. (2003). "Cancer mortality risk among workers at the Mayak nuclear facility," Radiat. Res. **159**, 787–798.

SHIMADA, Y., YASUKAWA-BARNES, J., KIM, R.Y., GOULD, M.N. and CLIFTON, K.H. (1994). "Age and radiation sensitivity of rat mammary clonogenic cells," Radiat. Res. **137**, 118–123.

SHIMIZU, Y., PIERCE, D.A., PRESTON, D.L. and MABUCHI, K. (1999). "Studies of the mortality of atomic bomb survivors. Report 12, Part II. Noncancer mortality: 1950–1990," Radiat. Res. **152**, 374–389.

SHIMOSATO, Y. (1989). "Pulmonary neoplasms," pages 785 to 827 in *Diagnostic Surgical Pathology*, Sternberg, S.S., Ed. (Raven Press, New York).

SHORE, R.E. (1990). "Overview of radiation-induced skin cancer in humans," Int. J. Radiat. Biol. **57**, 809–827.

SHORE, R.E. (1992). "Issues and epidemiological evidence regarding radiation-induced thyroid cancer," Radiat. Res. **131**, 98–111.
SHORE, R.E. (2001). "Radiation-induced skin cancer in humans," Med. Pediat. Oncol. **36**, 549–554.
SHORE, R.E., MOSESON, M., XUE, X., TSE, Y., HARLEY, N. and PASTERNAK, B.S. (2002). "Skin cancer after x-ray treatment for scalp ringworm," Radiat. Res. **157**, 410–418.
SILVER, A.R., BRECKON, G., MASSON, W.K., MALOWANY, D. and COX, R. (1987). "Studies on radiation myeloid leukemogenesis in the mouse," pages 494 to 500 in *Radiation Research, Proceedings of the 8th International Congress of Radiation Research*, Fielden, E.M., Fowler, J.F., Hendry, J.H. and Scott, D., Eds. (Taylor and Francis, New York).
SILVER, A.R., MASSON, W.K., BRECKON, G. and COX, R. (1988). "Preliminary molecular studies on two chromosome 2 encoded genes, c-*abl* and beta 2M, in radiation-induced murine myeloid leukaemias," Int. J. Radiat. Biol. Relat. Stud. Phys. Chem. Med. **53**, 57–63.
SILVER, A., MOODY, J., DUNFORD, R., CLARK, D., GANZ, S., BULMAN, R., BOUFFLER, S., FINNON, P., MEIJNE, E., HUISKAMP, R. and COX, R. (1999). "Molecular mapping of chromosome 2 deletions in murine radiation-induced AML localizes a putative tumor suppressor gene to a 1.0 cM region homologous to human chromosome segment 11p11-12," Genes Chromosomes Cancer **24**, 95–104.
SINCLAIR, W.K. (1963). "Absorbed dose in biological specimens irradiated externally with cobalt-60 gamma radiation," Radiat. Res. **20**, 288–297.
SLEBOS, R.J., KIBBELAAR, R.E., DALESIO, O., KOOISTRA, A., STAM, J., MEIJER, C.J., WAGENAAR, S.S., VANDERSCHUEREN, R.G., VAN ZANDWIJK, N., MOOI, W.J., BOS, J.L. and RODENHUIS, S. (1990). "K-*ras* oncogene activation as a prognostic marker in adenocarcinoma of the lung," N. Engl. J. Med. **323**, 561–565.
SLEBOS, R.J., HRUBAN, R.H., DALESIO, O., MOOI, W.J., OFFERHAUS, G.J. and RODENHUIS, S. (1991). "Relationship between K-*ras* oncogene activation and smoking in adenocarcinoma of the human lung," J. Natl. Cancer Inst. **83**, 1024–1027.
SMIDER, V., RATHMELL, W.K., LIEBER, M.R. and CHU, G. (1994). "Restoration of x-ray resistance and V(D)J recombination in mutant cells by Ku cDNA," Science **266**, 288–291.
SMITH, J.M. (1959). "A theory of aging," Nature **184**, 956–957.
SMITH, G.H. (1996). "Experimental mammary epithelial morphogenesis in an *in vivo* model: Evidence for distinct cellular progenitors of the ductal and lobular phenotypes," Breast Cancer Res. Treat. **39**, 21–31.
SMITH, G.H. and MEDINA, D. (1988). "A morphologically distinct candidate for an epithelial stem cell in mouse mammary gland," J. Cell Sci. **90**, 173–183.
SMITH, M.L., CHEN, I.T., ZHAN, Q., BAE, I., CHEN, C.Y., GILMER, T.M., KASTAN, M.B., O'CONNOR, P.M. and FORNACE, A.J., JR. (1994). "Interaction of the *p53*-regulated protein Gadd45 with proliferating cell nuclear antigen," Science **266**, 1376–1380.

SNEDECOR, G.W. (1946). *Statistical Methods Applied to Experiments in Agriculture and Biology*, 4th ed. (Iowa State University Press, Ames, Iowa).

SONTAG, S., DREW, C. and DREW, A.L. (1998). *Blind Man's Bluff: The Untold Story of American Submarine Espionage* (Public Affairs, New York).

SPENCER, W.P. and STERN, C. (1948). "Experiments to test the validity of the linear r-dose/mutation frequency relation in *Drosophila* at low dosage," Genetics **33**, 43–74.

SPIESS, H. (1969). "^{224}Ra-Induced Tumors in Children and Adults," pages 227 to 47 in *Delayed Effects of Bone-Seeking Radionuclides,*" Mays, C.W., Jee, W.S.S., Lloyd, R.D., Stover, B.J., Dougherty, J.H. and Taylor, G.N., Eds. (University of Utah Press, Salt Lake City, Utah).

SPIESS, H. (1995). "The ^{224}Ra study: Past, present and future," pages 157 to 164 in *Health Effects of Internally Deposited Radionuclides: Emphasis on Radium and Thorium*, van Kaick, G., Karaoglou, A. and Kellerer, A.M., Eds. (World Scientific, Hackensack, New Jersey).

SPIETHOFF, A., WESCH, H., HOVER, K.H. and WEGENER, K. (1992). "The combined and separate action of neutron radiation and zirconium dioxide on the liver of rats," Health Phys. **63**, 111–118.

STANLEY, L.A., DEVEREUX, T.R., FOLEY, J., LORD, P.G., MARONPOT, R.R., ORTON, T.C. and ANDERSON, M.W. (1992). "Proto-oncogene activation in liver tumors of hepatocarcinogenesis-resistant strains of mice," Carcinogenesis **13**, 2427–2433.

STANNARD, J.N. (1988). *Radioactivity and Health, a History*, DOE/RL/01830-T59 (National Technical Information Service, Springfield, Virginia).

STANTON, M.F. (1979). "Tumors of the bone," IARC Sci. Publ. **23**, 577–609.

STEGELMEIER, B.L., GILLETT, N.A., REBAR, A.H. and KELLY, G. (1991). "The molecular progression of plutonium-239-induced rat lung carcinogenesis: Ki-*ras* expression and activation," Mol. Carcinog. **4**, 43–51.

STENERLOW, B., HOGLUND, E., ELMROTH, K., KARLSSON, K.H. and RADULESCU, I. (2002). "Radiation quality dependence of DNA damage induction," Radiat. Prot. Dosim. **99**, 137–141.

STERN, R.S., THIBODEAU, L.A., KLEINERMAN, R.A., PARRISH, J.A. and FITZPATRICK, T.B. (1979). "Risk of cutaneous carcinoma in patients treated with oral methoxsalen photochemotherapy for psoriasis," N. Engl. J. Med. **300**, 809–813.

STOLER, D.L., CHEN, N., BASIK, M., KAHLENBERG, M.S., RODRIGUEZ-BIGAS, M.A., PETRELLI, N.J. and ANDERSON, G.R. (1999). "The onset and extent of genomic instability of sporadic colorectal tumor progression," Proc. Natl. Acad. Sci. USA **96**, 15121–15126.

STORER, J.B., MITCHELL, T.J. and FRY, R.J.M. (1988). "Extrapolation of the relative risk of radiogenic neoplasms across mouse strains and to man," Radiat. Res. **114**, 331–353.

STRASSER, A., HARRIS, A.W., JACKS, T. and CORY, S. (1994). "DNA damage can induce apoptosis in proliferating lymphoid cells via *p-53* independent mechanisms inhibitable by *Bcl-2*," Cell **79**, 329–339.

STREHLER, B.L. (1959). "Origin and comparison of the effects of time and high-energy radiations on living systems," Q. Rev. Biol. **34**, 117–142.

STREHLER, B.L. (1960). "Fluctuating energy demands as determinants of the death process (a parsimonious theory of the Gompertz function)," pages 309 to 314 in *The Biology of Aging*, Strehler, B.L., Ebert, J.D., Glass, H.B. and Shock, N.W., Eds. (American Institute of Biological Sciences, Washington).

STREHLER, B.L. and MILDVAN, A.S. (1960). "General theory of mortality and aging (a stochastic model relates observations on aging, physiologic decline, mortality, and radiation)," Science **132**, 14–21.

STURM, S.A., STRAUSS, P.G., ADOLPH, S., HAMEISTER, H. and ERFLE, V. (1990). "Amplification and rearrangement of c-*myc* in radiation-induced murine osteosarcomas," Cancer Res. **50**, 4146–4153.

SUTHERLAND, B.M., BENNETT, P.V., SIDORKINA, O. and LAVAL, J. (2000). "Clustered DNA damages induced in isolated DNA and in human cells by low doses of ionizing radiation," Proc. Natl. Acad. Sci. USA **97**, 103–108.

SUTHERLAND, B.M., BENNETT, P.V., SUTHERLAND, J.C. and LAVAL, J. (2002). "Clustered DNA damages induced by x rays in human cells," Radiat. Res. **157**, 611–616.

SWIFT, M., MORRELL, D., MASSEY, R.B. and CHASE, C.L. (1991). "Incidence of cancer in 161 families affected by ataxia-telangiectasia," N. Engl. J. Med. **325**, 1831–1836.

SZILARD, L. (1959). "On the nature of the aging process," Proc. Natl. Acad. Sci. USA **45**, 30–45.

SZOSTAK, J.W., ORR-WEAVER, T.L., ROTHSTEIN, R.J. and STAHL, F.W. (1983). "The double-strand-break repair model for recombination," Cell **33**, 25–35.

TACCIOLI, G.E., RATHBUN, G., OLTZ, E., STAMATO, T., JEGGO, P.A. and ALT, F.W. (1993). "Impairment of V(D)J recombination in double-strand break repair mutants," Science **260**, 207–210.

TACCIOLI, G.E., GOTTLIEB, T.M., BLUNT, T., PRIESTLEY, A., DEMENGEOT, J., MIZUTA, R., LEHMANN, A.R., ALT, F.W., JACKSON, S.P. and JEGGO, P.A. (1994). "Ku80: Product of the *XRCC5* gene and its role in DNA repair and V(D)J recombination," Science **265**, 1442–1445.

TANO, K., SHIOTA, S., COLLIER, J., FOOTE, R.S. and MITRA, S. (1990). "Isolation and structural characterization of a cDNA clone encoding the human DNA repair protein for O^6-alkylguanine," Proc. Natl. Acad. Sci. USA **87**, 686–690.

TAUCHI, H., KOBAYASHI, J., MORISHIMA, K., VAN GENT, D.C., SHIRAISHI, T., VERKAIK, N.S., VANHEEMS, D., ITO, E., NAKAMURA, A., SONODA, E., TAKATA, M., TAKEDA, S., MATSUURA, S. and KOMATSU, K. (2002). "*Nbs1* is essential for DNA repair by

homologous recombination in higher vertebrate cells," Nature **420**, 93–98.
TAYLOR, D.M., MAYS, C.W., GERBER, G.B. and THOMAS, R.G., Eds. (1989). *Risks from Radium and Thorotrast*, BIR Report 21 (British Institute of Radiology, London).
TAYLOR, G.N., LLOYD, R.D. and MAYS, C.W. (1993). "Liver cancer induction by ^{239}Pu, ^{241}Am, and thorotrast in the grasshopper mouse, Onychomys leukogaster," Health Phys. **64**, 141–146.
TAYLOR, J.A., WATSON, M.A., DEVEREUX, T.R., MICHELS, R.Y., SACCOMANNO, G. and ANDERSON, M. (1994). "*p53* mutation hotspot in radon-associated lung cancer," Lancet **343**, 86–87.
TAYLOR, G.N., LLOYD, R.D., MAYS, C.W., MILLER, S.C., JEE, W.S.S., MORI, S., SHABESTARI, L. and LI, X.J. (1997). "Relationship of natural incidence and radiosensitivity for bone cancer in dogs," Health Phys. **73**, 679–683.
TEMIN, H.M. (1985). "Genetic mechanisms of oncogenesis," pages 15 to 21 in *The Role of Chemicals and Radiation in the Etiology of Cancer, Carcinogenesis*, Huberman, E. and Barr, S.H., Eds. (Raven Press, New York).
TEOULE, R. (1987). "Radiation-induced DNA damage and its repair," Int. J. Radiat. Biol. Relat. Stud. Phys. Chem. Med. **51**, 573–589.
THACKER, J. and STRETCH, A. (1983). "Recovery from lethal and mutagenic damage during postirradiation holding and low-dose-rate irradiations of cultured hamster cells," Radiat. Res. **96**, 380–392.
THOMAS, R.G. (1995). "Tumorigenesis in the U.S. radium lumenizers: How unsafe was this occupation?," pages 145 to 149 in *Health Effects of Internally Deposited Radionuclides: Emphasis on Radium and Thorium*, van Kaick, G., Karaoglou, A. and Kellerer, A.M., Eds. (World Scientific, Hackensack, New Jersey).
THOMPSON, R.C., Ed. (1989). *Life-Span Effects of Ionizing Radiation in the Beagle Dog, Proceedings of the 22nd Hanford Symposium on Health and the Environment*, Report No. PNL-6822 (Battelle Press, Columbus, Ohio).
THOMPSON, L.H. and SCHILD, D. (2002). "Recombinational DNA repair and human disease," Mutat. Res. **509**, 49–78.
TOKUHATA, G.K. and LILIENFELD, A.M. (1963). "Familial aggregation of lung cancer in humans," J. Natl. Cancer Inst. **30**, 239.
TOKUNAGA, M., LAND, C.E., YAMAMOTO, T., ASANO, M., TOKUOKA, S., EZAKI, H., NISHIMORI, I. and FUJIKURA, T. (1984). "Breast cancer among bomb survivors," pages 45 to 56 in *Radiation Carcinogenesis: Epidemiology and Biological Significance*, Boice, J.D., Jr. and Fraumeni, J.F., Jr., Eds. (Raven Press, New York).
TOLLE, D.V., FRITZ, T.E. and SEED, T.M. (1997). "Radiation-induced leukemia," pages 33 to 73 in *Atlas of Experimentally-Induced Neoplasia in the Beagle Dog*, Watson, C.R. and Dagle, G.E., Eds. (Battelle Press, Columbus, Ohio).
TOLLE, D.V., FRITZ, T.E., SEED, T.M., CULLEN, S.M., LOMBARD, L.S. and POOLE, C.M. (1982). "Leukemia induction in beagles exposed

continuously to ^{60}Co gamma irradiation: Hematopathology," pages 241 to 249 in *Experimental Hematology Today 1982*, Baum, S.J., Ledney, G.D. and Thierfelder, S., Eds. (S. Karger, AG, Basel, Switzerland).

TRAKHTENBROT, L., KRAUTHGAMER, R., RESNITZKY, P. and HARAN-GHERA, N. (1988). "Deletion of chromosome 2 is an early event in the development of radiation-induced myeloid leukemia in SJL/J mice," Leukemia **2**, 545–550.

TRAVIS, C.C. and WHITE, R.K. (1988). "Interspecific scaling of toxicity data," Risk Anal. **8**, 119–125.

ULLRICH, R.L. and PRESTON, R.J. (1987). "Myeloid leukemia in male RFM mice following irradiation with fission spectrum neutrons or gamma rays," Radiat. Res. **109**, 165–170.

ULLRICH, R.L. and STORER, J.B. (1979). "Influence of gamma irradiation on the development of neoplastic disease in mice: III. Dose-rate effects," Radiat. Res. **80**, 325–342.

UNSCEAR (1958). United Nations Scientific Committee on the Effects of Atomic Radiation. "The genetic effects of radiation," pages 172 to 204 in *Report of the United Nations Scientific Committee on the Effects of Atomic Radiation General Assembly* (United Nations Publications, New York).

UNSCEAR (1982). United Nations Scientific Committee on the Effects of Atomic Radiation. *Ionizing Radiation: Sources and Biological Effects, 1982 Report to the General Assembly, with Annexes* (United Nations Publications, New York).

UNSCEAR (1986). United Nations Scientific Committee on the Effects of Atomic Radiation. *Genetic and Somatic Effects of Ionizing Radiation, 1986 Report to the General Assembly, with Annexes* (United Nations Publications, New York).

UNSCEAR (1988). United Nations Scientific Committee on the Effects of Atomic Radiation. "Annex E: Genetic hazards," pages 375 to 403 in *Sources, Effects and Risks of Ionizing Radiation. Annex E: Genetic Hazards, 1988 Report to the General Assembly, with Annexes* (United Nations Publications, New York).

UNSCEAR (1994). United Nations Scientific Committee on the Effects of Atomic Radiation. *Sources and Effects of Ionizing Radiation, 1994 Report to the General Assembly, with Scientific Annexes* (United Nations Publications, New York).

UNSCEAR (2000). United Nations Scientific Committee on the Effects of Atomic Radiation. *Sources and Effects of Ionizing Radiation, Volume II: Effects, 2000 Report to the General Assembly, with Scientific Annexes* (United Nations Publications, New York).

UNSCEAR (2001). United Nations Scientific Committee on the Effects of Atomic Radiation. *Hereditary Effects of Radiation, 2001 Report to the General Assembly with Scientific Annex* (United Nations Publications, New York).

UPTON, A.C. (1977). "Experimental radiation-induced leukemia," pages 37 to 50 in *Radiation-Induced Leukemogenesis and Related Viruses*,

Duplan, J.F., Ed. (Elsevier/North-Holland Publishing Company, New York).

UPTON, A.C. (2001). "Radiation hormesis: Data and interpretations," Crit. Rev. Toxicol. **31**, 681–695.

UPTON, A.C., WOLFF, F.F., FURTH, J. and KIMBALL, A.W. (1958). "A comparison of the induction of myeloid and lymphoid leukemias in x-radiated RF mice," Cancer Res. **18**, 842–848.

VAHAKANGAS, K.H., SAMET, J.M., METCALF, R.A., WELSH, J.A., BENNETT, W.P., LANE, D.P. and HARRIS, C.C. (1992). "Mutations of *p53* and *ras* genes in radon-associated lung cancer from uranium miners," Lancet **339**, 576–580.

VAN DER RAUWELAERT, E., MAISIN, J.R. and MERREGAERT, J. (1988). "Provirus integration and *myc* amplification in ^{90}Sr induced osteosarcomas of CF_1 mice," Oncogene **2**, 215–222.

VAN DER SCHANS, G.P. (1969). "Abstract: On the production of breaks in the DNA by x-rays," Int. J. Radiat. Biol. **16**, 58.

VAN DER SCHANS, G.P., PATERSON, M.C. and CROSS, W.G. (1983). "DNA strand break and rejoining in cultured human fibroblasts exposed to fast neutrons or gamma rays," Int. J. Radiat Biol. Relat. Stud. Phys. Chem. Med. **44**, 75–85.

VAN KRANEN, H.J., VERMEULEN, E., SCHOREN, L., BAX, J., WOUTERSEN, R.A., VAN IERSEL, P., VAN KREIJL, C.F. and SCHERER, E. (1991). "Activation of c-K-*ras* is frequent in pancreatic carcinomas of Syrian hamsters, but is absent in pancreatic tumors of rats," Carcinogenesis **12**, 1477–1482.

VAN LOHUIZEN, M., BREUR, M. and BERNS, A. (1989). "N-*myc* is frequently activated by proviral insertion in MuLV-induced T cell lymphomas," EMBO J. **8**, 133–136.

VASMEL, W.L.E., RADASZKIEWICZ, T., MILTENBURG, A.M., ZIJLSTRA M. and MELIEF, C.J.M. (1987). "Refinement and precision in the classification of murine lymphomas by genotyping with immunoglobulin and T cell receptor probes," Leukemia **1**, 155–162.

VAUGHAN, J.M. (1973). "Skeletal tumors induced by internal radiation," pages 377 to 394 in *Bone: Certain Aspects of Neoplasia*, Price, C.H.G. and Ross, F.G.M., Eds. (Butterworths, London).

WAGA, S., HANNON, G.J., BEACH, D. and STILLMAN, B. (1994). "The *p21* inhibitor of cyclin-dependent kinases controls DNA replication by interaction with PCNA," Nature **369**, 574–578.

WAHLS, W.P., WALLACE, L.J. and MOORE, P.D. (1990). "Hypervariable minisatellite DNA is a hotspot for homologous recombination in human cells," Cell **60**, 95–103.

WALKER, C., GOLDSWORTHY, T.L., WOLF, D.C. and EVERITT, J. (1992). "Predisposition to renal cell carcinoma due to alteration of a cancer susceptibility gene," Science **255**, 1693–1695.

WALLACE, S.S. (1988). "AP endonucleases and DNA glycosylases that recognize oxidative DNA damage," Environ. Mol. Mutagen. **12**, 431–477.

WARD, J.F. (1988). "DNA damage produced by ionizing radiation in mammalian cells: Identities, mechanisms of formation, and reparability," Prog. Nucleic Acid Res. Mol. Biol. **35**, 95–125.

WARD, J.F. (1994). "The complexity of DNA damage: Relevance to biological consequences," Int. J. Radiat. Biol. **66**, 427–432.

WATANABE, K., BOIS, F.Y. and ZEISE, L. (1992). "Interspecies extrapolation: A reexamination of acute toxicity data," Risk Anal. **12**, 301–310.

WATSON, J.D., ESZES, M., OVERELL, R., CONLON, P., WIDMER, M. and GILLIS, S. (1987). "Effect of infection with murine recombinant retroviruses containing the v-src oncogene on interleukin 2- and interleukin 3-dependent growth states," J. Immunol. **139**, 123–129.

WEISS, W. (1983). "The epidemiology of lung cancer," pages 1 to 8 in *Comparative Respiratory Tract Carcinogenesis*, Vol. 1, Reznik-Schuller, H.M., Ed. (CRC Press, Boca Raton, Florida).

WEISS, W. (1991). "Chronic obstructive pulmonary disease and lung cancer," in *Chronic Obstructive Pulmonary Disease*, 1st ed., Cherniak N.S., Ed. (Elsevier Science, New York).

WELLINGS, S.R. and JENSEN, H.M. (1973). "On the origin and progression of ducted carcinoma in the human breast," J. Natl. Cancer Inst. **50**, 1111–1118.

WESCH, H., VAN KAICK, G., RIEDEL, W., KAUL, A., WEGENER, K., HASENOHRL, K., IMMICH, H. and MUTH, H. (1983). "Recent results of the German thorotrast study—statistical evaluation of animal experiments with regard to the nonradiation effects in human thorotrastosis," Health Phys. **44** (Suppl. 1), 317–321.

WEST, M., BLANCHETTE, C., DRESSMAN, H., HUANG, E., ISHIDA, S., SPANG, R., ZUZAN, H., OLSON, J.A., JR., MARKS, J.R. and NEVINS, J.R. (2001). "Predicting the clinical status of human breast cancer by using gene expression profiles," Proc. Nat. Acad. Sci. USA **98**, 11462–11467.

WHITAKER, S.J., POWELL, S.N. and MCMILLAN, T.J. (1991). "Molecular assays of radiation-induced DNA damage," Eur. J. Cancer **27**, 922–928.

WHO (1981). World Health Organization. *Histological Typing of Lung Tumours, International Histological Classification of Tumours*, No. 1, 2nd ed. (World Health Organization, Geneva).

WHO (2000). World Health Organization. *International Classification of Diseases for Oncology*, 3rd ed. (World Health Organization, Geneva).

WICK, R.R., NEKOLLA, E.A., GOSSNER, W. and KELLERER, A.M. (1999). "Late effects in ankylosing spondylitis patients treated with ^{224}Ra," Radiat. Res. **152** (Suppl. 6), S8–S11.

WIENER, F. and POTTER, M. (1993). "*myc*-associated chromosomal translocations and rearrangements in plasmacytomagenesis," pages 91 to 124 in *The Causes and Consequences of Chromosomal Aberrations*, Kirsch, I.R. Ed. (CRC Press, Boca Raton, Florida).

WIENER, F., BABONITS, M., SPIRA, J., KLEIN, G. and BAZIN, H. (1982). "Non-random chromosomal changes involving chromosomes 6 and 7 in spontaneous rat immunocytomas," Int. J. Cancer **29**, 431–437.

WILLIAMS, G.C. (1957). "Pleiotropy, natural selection, and the evolution of senescence," Evolution **11**, 398–411.

WITSCHI, H. and SCHULLER, H.M. (1991). "Diffuse and continuous cell proliferation enhances radiation-induced tumorigenesis in hamster lung," Cancer Lett. **60**, 193–197.

WONG, A.J., RUPPERT, J.M., EGGLESTON, J., HAMILTON, S.R., BAYLIN, S.B. and VOGELSTEIN, B. (1986). "Gene amplification of c-*myc* and N-*myc* in small cell carcinoma of the lung," Science **233**, 461–464.

WOOD, R.D. (1995). "Proteins that participate in nucleotide excision repair of DNA in mammalian cells," Philos. Trans. R. Soc. Lond. B. Biol. Sci. **347**, 69–74.

WOOD, R.D., MITCHELL, M., SGOUROS, J. and LINDAHL, T. (2001). "Human DNA repair genes," Science **291**, 1284–1289.

WRIGHT, S. (1950). "Discussion of population genetics and radiation," J. Cell Physiol. **35** (Suppl. 1), 187–205.

WYNDER, E.L. and COVEY, L.S. (1987). "Epidemiologic patterns in lung cancer by histologic type," Eur. J. Cancer Clin. Oncol. **23**, 1491–1496.

XIONG, Y., HANNON, G.J., ZHANG, H., CASSO, D., KOBAYASHI, R. and BEACH, D. (1993). "*p21* is a universal inhibitor of cyclin kinases," Nature **366**, 701–704.

YOU, M., CANDRIAN, U., MARONPOT, R.R., STONER, G.D. and ANDERSON, M.W. (1989). "Activation of the Ki-*ras* protooncogene in spontaneously occurring and chemically induced lung tumors of the strain A mouse," Proc. Natl. Acad. Sci. USA **86**, 3070–3074.

YOU, M., WANG, Y., LINEEN, A., STONER, G.D., YOU, L.A., MARONPOT, R.R. and ANDERSON, M.W. (1991). "Activation of protooncogenes in mouse lung tumors," Exp. Lung Res. **17**, 389–400.

ZAHARKO, D.S. (1974). "Pharmacokinetics and drug effect," Biochem Pharmacol. **23** (Suppl. 2), 1–8.

ZARBL, H., SUKUMAR, S., ARTHUR, A.V., MARTIN-ZANCA, D. and BARBACID, M. (1985). "Direct mutagenesis of Ha-*ras*-1 oncogenes by N-nitroso-N-methylurea during initiation of mammary carcinogenesis in rats," Nature **315**, 382–385.

ZHANG, R., HAAG, J.D. and GOULD, M.N. (1991). "Quantitating the frequency of initiation and *cH-ras* mutation in *in situ* N-methyl-N-nitrosourea-exposed rat mammary gland," Cell Growth Diff. **2**, 1–6.

ZHANG, N., CHEN, P., GATEI, M., SCOTT, S., KHANNA, K.K. and LAVIN, M.F. (1998). "An anti-sense construct of full-length ATM cDNA imposes a radiosensitive phenotype on normal cells," Oncogene **17**, 811–818.

The NCRP

The National Council on Radiation Protection and Measurements is a nonprofit corporation chartered by Congress in 1964 to:

1. Collect, analyze, develop and disseminate in the public interest information and recommendations about (a) protection against radiation and (b) radiation measurements, quantities and units, particularly those concerned with radiation protection.
2. Provide a means by which organizations concerned with the scientific and related aspects of radiation protection and of radiation quantities, units and measurements may cooperate for effective utilization of their combined resources, and to stimulate the work of such organizations.
3. Develop basic concepts about radiation quantities, units and measurements, about the application of these concepts, and about radiation protection.
4. Cooperate with the International Commission on Radiological Protection, the International Commission on Radiation Units and Measurements, and other national and international organizations, governmental and private, concerned with radiation quantities, units and measurements and with radiation protection.

The Council is the successor to the unincorporated association of scientists known as the National Committee on Radiation Protection and Measurements and was formed to carry on the work begun by the Committee in 1929.

The participants in the Council's work are the Council members and members of scientific and administrative committees. Council members are selected solely on the basis of their scientific expertise and serve as individuals, not as representatives of any particular organization. The scientific committees, composed of experts having detailed knowledge and competence in the particular area of the committee's interest, draft proposed recommendations. These are then submitted to the full membership of the Council for careful review and approval before being published.

The following comprise the current officers and membership of the Council:

Officers

President	Thomas S. Tenforde
Senior Vice President	Kenneth R. Kase
Secretary and Treasurer	David A. Schauer
Assistant Secretary	Michael F. McBride

THE NCRP / 243

Members

John F. Ahearne
Sally A. Amundson
Larry E. Anderson
Benjamin R. Archer
Mary M. Austin-Seymour
Steven M. Becker
Joel S. Bedford
Eleanor A. Blakely
William F. Blakely
John D. Boice, Jr.
Wesley E. Bolch
Thomas B. Borak
Andre Bouville
Leslie A. Braby
David J. Brenner
James A. Brink
Antone L. Brooks
Jerrold T. Bushberg
John F. Cardella
Stephanie K. Carlson
Polly Y. Chang
S.Y. Chen
Kelly L. Classic
Mary E. Clark
James E. Cleaver
Michael L. Corradini
J. Donald Cossairt
Allen G. Croff
Francis A. Cucinotta
Paul M. DeLuca
John F. Dicello, Jr.
William P. Dornsife
David A. Eastmond

Stephen A. Feig
Kenneth R. Foster
John R. Frazier
Donald P. Frush
Thomas F. Gesell
Andrew J. Grosovsky
Raymond A. Guilmette
Roger W. Harms
John W. Hirshfeld, Jr.
F. Owen Hoffman
Roger W. Howell
Kenneth R. Kase
Ann R. Kennedy
William E. Kennedy, Jr.
David C. Kocher
Ritsuko Komaki
Amy Kronenberg
Susan M. Langhorst
Howard L. Liber
James C. Lin
Jill A. Lipoti
John B. Little
Paul A. Locke
Jay H. Lubin
C. Douglas Maynard
Claire M. Mays
Cynthia H. McCollough
Barbara J. McNeil
Fred A. Mettler, Jr.
Charles W. Miller
Jack Miller
Kenneth L. Miller
William H. Miller
William F. Morgan

John E. Moulder
David S. Myers
Bruce A. Napier
Carl J. Paperiello
R. Julian Preston
Allan C.B. Richardson
Henry D. Royal
Marvin Rosenstein
Michael T. Ryan
Jonathan M. Samet
Thomas M. Seed
Stephen M. Seltzer
Roy E. Shore
Edward A. Sickles
Steven L. Simon
Paul Slovic
Christopher G. Soares
Daniel J. Strom
Thomas S. Tenforde
Julie E.K. Timins
Lawrence W. Townsend
Lois B. Travis
Robert L. Ullrich
Richard J. Vetter
Daniel E. Wartenberg
Chris G. Whipple
Stuart C. White
J. Frank Wilson
Susan D. Wiltshire
Gayle E. Woloschak
Shiao Y. Woo
Marco A. Zaider
Pasquale D. Zanzonico

Honorary Members

Warren K. Sinclair, *President Emeritus;* Charles B. Meinhold, *President Emeritus*
S. James Adelstein, *Honorary Vice President*
W. Roger Ney, *Executive Director Emeritus*
William M. Beckner, *Executive Director Emeritus*

Seymour Abrahamson
Edward L. Alpen
Lynn R. Anspaugh
John A. Auxier
William J. Bair
Harold L. Beck
Bruce B. Boecker
Victor P. Bond
Robert L. Brent
Reynold F. Brown
Melvin C. Carter
Randall S. Caswell
Frederick P. Cowan
James F. Crow
Gerald D. Dodd
Sarah S. Donaldson

Patricia W. Durbin
Keith F. Eckerman
Thomas S. Ely
Richard F. Foster
R.J. Michael Fry
Ethel S. Gilbert
Joel E. Gray
Robert O. Gorson
Arthur W. Guy
Eric J. Hall
Naomi H. Harley
William R. Hendee
Donald G. Jacobs
Bernd Kahn
Charles E. Land
Roger O. McClellan

Dade W. Moeller
A. Alan Moghissi
Wesley L. Nyborg
John W. Poston, Sr.
Andrew K. Poznanski
Chester R. Richmond
Genevieve S. Roessler
Lawrence N. Rothenberg
Eugene L. Saenger
William J. Schull
John B. Storer
John E. Till
Arthur C. Upton
Edward W. Webster
F. Ward Whicker
Marvin C. Ziskin

Lauriston S. Taylor Lecturers

John B. Little (2005) *Nontargeted Effects of Radiation: Implications for Low-Dose Exposures*
Abel J. Gonzalez (2004) *Radiation Protection in the Aftermath of a Terrorist Attack Involving Exposure to Ionizing Radiation*
Charles B. Meinhold (2003) *The Evolution of Radiation Protection: From Erythema to Genetic Risks to Risks of Cancer to ?*
R. Julian Preston (2002) *Developing Mechanistic Data for Incorporation into Cancer Risk Assessment: Old Problems and New Approaches*
Wesley L. Nyborg (2001) *Assuring the Safety of Medical Diagnostic Ultrasound*
S. James Adelstein (2000) *Administered Radioactivity: Unde Venimus Quoque Imus*
Naomi H. Harley (1999) *Back to Background*
Eric J. Hall (1998) *From Chimney Sweeps to Astronauts: Cancer Risks in the Workplace*
William J. Bair (1997) *Radionuclides in the Body: Meeting the Challenge!*
Seymour Abrahamson (1996) *70 Years of Radiation Genetics: Fruit Flies, Mice and Humans*
Albrecht Kellerer (1995) *Certainty and Uncertainty in Radiation Protection*
R.J. Michael Fry (1994) *Mice, Myths and Men*
Warren K. Sinclair (1993) *Science, Radiation Protection and the NCRP*
Edward W. Webster (1992) *Dose and Risk in Diagnostic Radiology: How Big? How Little?*
Victor P. Bond (1991) *When is a Dose Not a Dose?*
J. Newell Stannard (1990) *Radiation Protection and the Internal Emitter Saga*
Arthur C. Upton (1989) *Radiobiology and Radiation Protection: The Past Century and Prospects for the Future*
Bo Lindell (1988) *How Safe is Safe Enough?*
Seymour Jablon (1987) *How to be Quantitative about Radiation Risk Estimates*
Herman P. Schwan (1986) *Biological Effects of Non-ionizing Radiations: Cellular Properties and Interactions*
John H. Harley (1985) *Truth (and Beauty) in Radiation Measurement*
Harald H. Rossi (1984) *Limitation and Assessment in Radiation Protection*
Merril Eisenbud (1983) *The Human Environment—Past, Present and Future*
Eugene L. Saenger (1982) *Ethics, Trade-Offs and Medical Radiation*
James F. Crow (1981) *How Well Can We Assess Genetic Risk? Not Very*
Harold O. Wyckoff (1980) *From "Quantity of Radiation" and "Dose" to "Exposure" and "Absorbed Dose"—An Historical Review*
Hymer L. Friedell (1979) *Radiation Protection—Concepts and Trade Offs*
Sir Edward Pochin (1978) *Why be Quantitative about Radiation Risk Estimates?*
Herbert M. Parker (1977) *The Squares of the Natural Numbers in Radiation Protection*

Currently, the following committees are actively engaged in formulating recommendations:

Program Area Committee 1: Basic Criteria, Epidemiology, Radiobiology, and Risk
  SC 1-7 Information Needed to Make Radiation Protection Recommendations for Travel Beyond Low-Earth Orbit
  SC 1-8 Risk to Thyroid from Ionizing Radiation
  SC 1-13 Impact of Individual Susceptibility and Previous Radiation Exposure on Radiation Risk for Astronauts
  SC 1-15 Radiation Safety in NASA Lunar Missions
  SC 85 Risk of Lung Cancer from Radon
Program Area Committee 2: Operational Radiation Safety
  SC 2-1 Radiation Protection Recommendations for First Responders
  SC 46-13 Design of Facilities for Medical Radiation Therapy
  SC 46-17 Radiation Protection in Educational Institutions
Program Area Committee 3: Nonionizing Radiation
  SC 89-5 Study and Critical Evaluation of Radiofrequency Exposure Guidelines
Program Area Committee 4: Radiation Protection in Medicine
  SC 4-1 Management of Persons Contaminated with Radionuclides
  SC 91-1 Precautions in the Management of Patients Who Have Received Therapeutic Amounts of Radionuclides
Program Area Committee 5: Environmental Radiation and Radioactive Waste Issues
  SC 64-22 Design of Effective Effluent and Environmental Monitoring Programs
  SC 64-23 Cesium in the Environment
  SC 87-3 Performance Assessment of Near Surface Radioactive Waste Facilities
Program Area Committee 6: Radiation Measurements and Dosimetry
  SC 6-1 Uncertainties in the Measurement and Dosimetry of External Radiation Sources
  SC 6-2 Radiation Exposure of the U.S. Population
  SC 6-3 Uncertainties in Internal Radiation Dosimetry
  SC 57-17 Radionuclide Dosimetry Models for Wounds
Advisory Committee 1: Public Policy and Risk Communication

In recognition of its responsibility to facilitate and stimulate cooperation among organizations concerned with the scientific and related aspects of radiation protection and measurement, the Council has created a category of NCRP Collaborating Organizations. Organizations or groups of organizations that are national or international in scope and are concerned with scientific problems involving radiation quantities, units, measurements and effects, or radiation protection may be admitted to collaborating status by the Council. Collaborating Organizations provide a means by which NCRP can gain input into its activities from a wider segment of society. At the same time, the relationships with the Collaborating Organizations facilitate wider dissemination of information about the Council's activities, interests and concerns. Collaborating Organizations have the opportunity to comment on draft reports (at the time that these are submitted to the members of the Council). This is intended to capitalize on the fact that Collaborating Organizations are in an excellent position to both contribute to the identification of what needs to be treated in NCRP reports and to identify problems that might result from proposed recommendations. The present Collaborating Organizations with which NCRP maintains liaison are as follows:

American Academy of Dermatology
American Academy of Environmental Engineers
American Academy of Health Physics
American Association of Physicists in Medicine
American College of Medical Physics
American College of Nuclear Physicians
American College of Occupational and Environmental Medicine
American College of Radiology
American Conference of Governmental Industrial Hygienists
American Dental Association
American Industrial Hygiene Association
American Institute of Ultrasound in Medicine
American Medical Association
American Nuclear Society
American Pharmaceutical Association
American Podiatric Medical Association
American Public Health Association
American Radium Society
American Roentgen Ray Society
American Society for Therapeutic Radiology and Oncology
American Society of Emergency Radiology
American Society of Health-System Pharmacists
American Society of Radiologic Technologists
Association of Educators in Radiological Sciences, Inc.
Association of University Radiologists
Bioelectromagnetics Society
Campus Radiation Safety Officers
College of American Pathologists
Conference of Radiation Control Program Directors, Inc.
Council on Radionuclides and Radiopharmaceuticals
Defense Threat Reduction Agency
Electric Power Research Institute
Federal Communications Commission
Federal Emergency Management Agency

Genetics Society of America
Health Physics Society
Institute of Electrical and Electronics Engineers, Inc.
Institute of Nuclear Power Operations
International Brotherhood of Electrical Workers
National Aeronautics and Space Administration
National Association of Environmental Professionals
National Center for Environmental Health/Agency for Toxic Substances
National Electrical Manufacturers Association
National Institute for Occupational Safety and Health
National Institute of Standards and Technology
Nuclear Energy Institute
Office of Science and Technology Policy
Paper, Allied-Industrial, Chemical and Energy Workers International Union
Product Stewardship Institute
Radiation Research Society
Radiological Society of North America
Society for Risk Analysis
Society of Chairmen of Academic Radiology Departments
Society of Nuclear Medicine
Society of Radiologists in Ultrasound
Society of Skeletal Radiology
U.S. Air Force
U.S. Army
U.S. Coast Guard
U.S. Department of Energy
U.S. Department of Housing and Urban Development
U.S. Department of Labor
U.S. Department of Transportation
U.S. Environmental Protection Agency
U.S. Navy
U.S. Nuclear Regulatory Commission
U.S. Public Health Service
Utility Workers Union of America

NCRP has found its relationships with these organizations to be extremely valuable to continued progress in its program.

Another aspect of the cooperative efforts of NCRP relates to the Special Liaison relationships established with various governmental organizations that have an interest in radiation protection and measurements. This liaison relationship provides: (1) an opportunity for participating organizations to designate an individual to provide liaison between the organization and NCRP; (2) that the individual designated will receive copies of draft NCRP reports (at the time that these are submitted to the members of the Council) with an invitation to comment, but not vote; and

(3) that new NCRP efforts might be discussed with liaison individuals as appropriate, so that they might have an opportunity to make suggestions on new studies and related matters. The following organizations participate in the Special Liaison Program:

Australian Radiation Laboratory
Bundesamt fur Strahlenschutz (Germany)
Canadian Nuclear Safety Commission
Central Laboratory for Radiological Protection (Poland)
China Institute for Radiation Protection
Commonwealth Scientific Instrumentation Research
  Organization (Australia)
European Commission
Health Council of the Netherlands
Institut de Radioprotection et de Surete Nucleaire
International Commission on Non-ionizing Radiation Protection
International Commission on Radiation Units and Measurements
Japan Radiation Council
Korea Institute of Nuclear Safety
National Radiological Protection Board (United Kingdom)
Russian Scientific Commission on Radiation Protection
South African Forum for Radiation Protection
World Association of Nuclear Operations
World Health Organization, Radiation and Environmental Health

NCRP values highly the participation of these organizations in the Special Liaison Program.

The Council also benefits significantly from the relationships established pursuant to the Corporate Sponsor's Program. The program facilitates the interchange of information and ideas and corporate sponsors provide valuable fiscal support for the Council's program. This developing program currently includes the following Corporate Sponsors:

3M
Duke Energy Corporation
GE Healthcare
Global Dosimetry Solutions
Landauer, Inc.
Nuclear Energy Institute
Southern California Edison Company

The Council's activities have been made possible by the voluntary contribution of time and effort by its members and participants and the generous support of the following organizations:

3M Health Physics Services
Agfa Corporation

Alfred P. Sloan Foundation
Alliance of American Insurers
American Academy of Dermatology
American Academy of Health Physics
American Academy of Oral and Maxillofacial Radiology
American Association of Physicists in Medicine
American Cancer Society
American College of Medical Physics
American College of Nuclear Physicians
American College of Occupational and Environmental Medicine
American College of Radiology
American College of Radiology Foundation
American Dental Association
American Healthcare Radiology Administrators
American Industrial Hygiene Association
American Insurance Services Group
American Medical Association
American Nuclear Society
American Osteopathic College of Radiology
American Podiatric Medical Association
American Public Health Association
American Radium Society
American Roentgen Ray Society
American Society of Radiologic Technologists
American Society for Therapeutic Radiology and Oncology
American Veterinary Medical Association
American Veterinary Radiology Society
Association of Educators in Radiological Sciences, Inc.
Association of University Radiologists
Battelle Memorial Institute
Canberra Industries, Inc.
Chem Nuclear Systems
Center for Devices and Radiological Health
College of American Pathologists
Committee on Interagency Radiation Research
 and Policy Coordination
Commonwealth Edison
Commonwealth of Pennsylvania
Consolidated Edison
Consumers Power Company
Council on Radionuclides and Radiopharmaceuticals
Defense Nuclear Agency
Defense Threat Reduction Agency
Eastman Kodak Company
Edison Electric Institute
Edward Mallinckrodt, Jr. Foundation
EG&G Idaho, Inc.

Electric Power Research Institute
Electromagnetic Energy Association
Federal Emergency Management Agency
Florida Institute of Phosphate Research
Florida Power Corporation
Fuji Medical Systems, U.S.A., Inc.
Genetics Society of America
Global Dosimetry Solutions
Health Effects Research Foundation (Japan)
Health Physics Society
ICN Biomedicals, Inc.
Institute of Nuclear Power Operations
James Picker Foundation
Martin Marietta Corporation
Motorola Foundation
National Aeronautics and Space Administration
National Association of Photographic Manufacturers
National Cancer Institute
National Electrical Manufacturers Association
National Institute of Standards and Technology
New York Power Authority
Philips Medical Systems
Picker International
Public Service Electric and Gas Company
Radiation Research Society
Radiological Society of North America
Richard Lounsbery Foundation
Sandia National Laboratory
Siemens Medical Systems, Inc.
Society of Nuclear Medicine
Society of Pediatric Radiology
U.S. Department of Energy
U.S. Department of Labor
U.S. Environmental Protection Agency
U.S. Navy
U.S. Nuclear Regulatory Commission
Victoreen, Inc.
Westinghouse Electric Corporation

Initial funds for publication of NCRP reports were provided by a grant from the James Picker Foundation.

NCRP seeks to promulgate information and recommendations based on leading scientific judgment on matters of radiation protection and measurement and to foster cooperation among organizations concerned with these matters. These efforts are intended to serve the public interest and the Council welcomes comments and suggestions on its reports or activities.

NCRP Publications

NCRP publications can be obtained online in both hard- and soft-copy (downloadable PDF) formats at http://NCRPpublications.org. Professional societies can arrange for discounts for their members by contacting NCRP. Additional information on NCRP publications may be obtained from the NCRP website (http://NCRPonline.org) or by telephone (800-229-2652, ext. 25) and fax (301-907-8768). The mailing address is:

> NCRP Publications
> 7910 Woodmont Avenue
> Suite 400
> Bethesda, MD 20814-3095

Abstracts of NCRP reports published since 1980, abstracts of all NCRP commentaries, and the text of all NCRP statements are available at the NCRP website. Currently available publications are listed below.

NCRP Reports

No. Title

8 *Control and Removal of Radioactive Contamination in Laboratories* (1951)
22 *Maximum Permissible Body Burdens and Maximum Permissible Concentrations of Radionuclides in Air and in Water for Occupational Exposure* (1959) [includes Addendum 1 issued in August 1963]
25 *Measurement of Absorbed Dose of Neutrons, and of Mixtures of Neutrons and Gamma Rays* (1961)
27 *Stopping Powers for Use with Cavity Chambers* (1961)
30 *Safe Handling of Radioactive Materials* (1964)
32 *Radiation Protection in Educational Institutions* (1966)
35 *Dental X-Ray Protection* (1970)
36 *Radiation Protection in Veterinary Medicine* (1970)
37 *Precautions in the Management of Patients Who Have Received Therapeutic Amounts of Radionuclides* (1970)
38 *Protection Against Neutron Radiation* (1971)
40 *Protection Against Radiation from Brachytherapy Sources* (1972)
41 *Specification of Gamma-Ray Brachytherapy Sources* (1974)
42 *Radiological Factors Affecting Decision-Making in a Nuclear Attack* (1974)

44 *Krypton-85 in the Atmosphere—Accumulation, Biological Significance, and Control Technology* (1975)
46 *Alpha-Emitting Particles in Lungs* (1975)
47 *Tritium Measurement Techniques* (1976)
49 *Structural Shielding Design and Evaluation for Medical Use of X Rays and Gamma Rays of Energies Up to 10 MeV* (1976)
50 *Environmental Radiation Measurements* (1976)
52 *Cesium-137 from the Environment to Man: Metabolism and Dose* (1977)
54 *Medical Radiation Exposure of Pregnant and Potentially Pregnant Women* (1977)
55 *Protection of the Thyroid Gland in the Event of Releases of Radioiodine* (1977)
57 *Instrumentation and Monitoring Methods for Radiation Protection* (1978)
58 *A Handbook of Radioactivity Measurements Procedures*, 2nd ed. (1985)
60 *Physical, Chemical, and Biological Properties of Radiocerium Relevant to Radiation Protection Guidelines* (1978)
61 *Radiation Safety Training Criteria for Industrial Radiography* (1978)
62 *Tritium in the Environment* (1979)
63 *Tritium and Other Radionuclide Labeled Organic Compounds Incorporated in Genetic Material* (1979)
64 *Influence of Dose and Its Distribution in Time on Dose-Response Relationships for Low-LET Radiations* (1980)
65 *Management of Persons Accidentally Contaminated with Radionuclides* (1980)
67 *Radiofrequency Electromagnetic Fields—Properties, Quantities and Units, Biophysical Interaction, and Measurements* (1981)
68 *Radiation Protection in Pediatric Radiology* (1981)
69 *Dosimetry of X-Ray and Gamma-Ray Beams for Radiation Therapy in the Energy Range 10 keV to 50 MeV* (1981)
70 *Nuclear Medicine—Factors Influencing the Choice and Use of Radionuclides in Diagnosis and Therapy* (1982)
72 *Radiation Protection and Measurement for Low-Voltage Neutron Generators* (1983)
73 *Protection in Nuclear Medicine and Ultrasound Diagnostic Procedures in Children* (1983)
74 *Biological Effects of Ultrasound: Mechanisms and Clinical Implications* (1983)
75 *Iodine-129: Evaluation of Releases from Nuclear Power Generation* (1983)
76 *Radiological Assessment: Predicting the Transport, Bioaccumulation, and Uptake by Man of Radionuclides Released to the Environment* (1984)

77 *Exposures from the Uranium Series with Emphasis on Radon and Its Daughters* (1984)
78 *Evaluation of Occupational and Environmental Exposures to Radon and Radon Daughters in the United States* (1984)
79 *Neutron Contamination from Medical Electron Accelerators* (1984)
80 *Induction of Thyroid Cancer by Ionizing Radiation* (1985)
81 *Carbon-14 in the Environment* (1985)
82 *SI Units in Radiation Protection and Measurements* (1985)
83 *The Experimental Basis for Absorbed-Dose Calculations in Medical Uses of Radionuclides* (1985)
84 *General Concepts for the Dosimetry of Internally Deposited Radionuclides* (1985)
86 *Biological Effects and Exposure Criteria for Radiofrequency Electromagnetic Fields* (1986)
87 *Use of Bioassay Procedures for Assessment of Internal Radionuclide Deposition* (1987)
88 *Radiation Alarms and Access Control Systems* (1986)
89 *Genetic Effects from Internally Deposited Radionuclides* (1987)
90 *Neptunium: Radiation Protection Guidelines* (1988)
92 *Public Radiation Exposure from Nuclear Power Generation in the United States* (1987)
93 *Ionizing Radiation Exposure of the Population of the United States* (1987)
94 *Exposure of the Population in the United States and Canada from Natural Background Radiation* (1987)
95 *Radiation Exposure of the U.S. Population from Consumer Products and Miscellaneous Sources* (1987)
96 *Comparative Carcinogenicity of Ionizing Radiation and Chemicals* (1989)
97 *Measurement of Radon and Radon Daughters in Air* (1988)
99 *Quality Assurance for Diagnostic Imaging* (1988)
100 *Exposure of the U.S. Population from Diagnostic Medical Radiation* (1989)
101 *Exposure of the U.S. Population from Occupational Radiation* (1989)
102 *Medical X-Ray, Electron Beam and Gamma-Ray Protection for Energies Up to 50 MeV (Equipment Design, Performance and Use)* (1989)
103 *Control of Radon in Houses* (1989)
104 *The Relative Biological Effectiveness of Radiations of Different Quality* (1990)
105 *Radiation Protection for Medical and Allied Health Personnel* (1989)
106 *Limit for Exposure to "Hot Particles" on the Skin* (1989)
107 *Implementation of the Principle of As Low As Reasonably Achievable (ALARA) for Medical and Dental Personnel* (1990)

108 *Conceptual Basis for Calculations of Absorbed-Dose Distributions* (1991)
109 *Effects of Ionizing Radiation on Aquatic Organisms* (1991)
110 *Some Aspects of Strontium Radiobiology* (1991)
111 *Developing Radiation Emergency Plans for Academic, Medical or Industrial Facilities* (1991)
112 *Calibration of Survey Instruments Used in Radiation Protection for the Assessment of Ionizing Radiation Fields and Radioactive Surface Contamination* (1991)
113 *Exposure Criteria for Medical Diagnostic Ultrasound: I. Criteria Based on Thermal Mechanisms* (1992)
114 *Maintaining Radiation Protection Records* (1992)
115 *Risk Estimates for Radiation Protection* (1993)
116 *Limitation of Exposure to Ionizing Radiation* (1993)
117 *Research Needs for Radiation Protection* (1993)
118 *Radiation Protection in the Mineral Extraction Industry* (1993)
119 *A Practical Guide to the Determination of Human Exposure to Radiofrequency Fields* (1993)
120 *Dose Control at Nuclear Power Plants* (1994)
121 *Principles and Application of Collective Dose in Radiation Protection* (1995)
122 *Use of Personal Monitors to Estimate Effective Dose Equivalent and Effective Dose to Workers for External Exposure to Low-LET Radiation* (1995)
123 *Screening Models for Releases of Radionuclides to Atmosphere, Surface Water, and Ground* (1996)
124 *Sources and Magnitude of Occupational and Public Exposures from Nuclear Medicine Procedures* (1996)
125 *Deposition, Retention and Dosimetry of Inhaled Radioactive Substances* (1997)
126 *Uncertainties in Fatal Cancer Risk Estimates Used in Radiation Protection* (1997)
127 *Operational Radiation Safety Program* (1998)
128 *Radionuclide Exposure of the Embryo/Fetus* (1998)
129 *Recommended Screening Limits for Contaminated Surface Soil and Review of Factors Relevant to Site-Specific Studies* (1999)
130 *Biological Effects and Exposure Limits for "Hot Particles"* (1999)
131 *Scientific Basis for Evaluating the Risks to Populations from Space Applications of Plutonium* (2001)
132 *Radiation Protection Guidance for Activities in Low-Earth Orbit* (2000)
133 *Radiation Protection for Procedures Performed Outside the Radiology Department* (2000)
134 *Operational Radiation Safety Training* (2000)
135 *Liver Cancer Risk from Internally-Deposited Radionuclides* (2001)
136 *Evaluation of the Linear-Nonthreshold Dose-Response Model for Ionizing Radiation* (2001)

137 *Fluence-Based and Microdosimetric Event-Based Methods for Radiation Protection in Space* (2001)
138 *Management of Terrorist Events Involving Radioactive Material* (2001)
139 *Risk-Based Classification of Radioactive and Hazardous Chemical Wastes* (2002)
140 *Exposure Criteria for Medical Diagnostic Ultrasound: II. Criteria Based on all Known Mechanisms* (2002)
141 *Managing Potentially Radioactive Scrap Metal* (2002)
142 *Operational Radiation Safety Program for Astronauts in Low-Earth Orbit: A Basic Framework* (2002)
143 *Management Techniques for Laboratories and Other Small Institutional Generators to Minimize Off-Site Disposal of Low-Level Radioactive Waste* (2003)
144 *Radiation Protection for Particle Accelerator Facilities* (2003)
145 *Radiation Protection in Dentistry* (2003)
146 *Approaches to Risk Management in Remediation of Radioactively Contaminated Sites* (2004)
147 *Structural Shielding Design for Medical X-Ray Imaging Facilities* (2004)
148 *Radiation Protection in Veterinary Medicine* (2004)
149 *A Guide to Mammography and Other Breast Imaging Procedures* (2004)
150 *Extrapolation of Radiation-Induced Cancer Risks from Nonhuman Experimental Systems to Humans* (2005)

Binders for NCRP reports are available. Two sizes make it possible to collect into small binders the "old series" of reports (NCRP Reports Nos. 8–30) and into large binders the more recent publications (NCRP Reports Nos. 32–150). Each binder will accommodate from five to seven reports. The binders carry the identification "NCRP Reports" and come with label holders which permit the user to attach labels showing the reports contained in each binder.

The following bound sets of NCRP reports are also available:

Volume I. NCRP Reports Nos. 8, 22
Volume II. NCRP Reports Nos. 23, 25, 27, 30
Volume III. NCRP Reports Nos. 32, 35, 36, 37
Volume IV. NCRP Reports Nos. 38, 40, 41
Volume V. NCRP Reports Nos. 42, 44, 46
Volume VI. NCRP Reports Nos. 47, 49, 50, 51
Volume VII. NCRP Reports Nos. 52, 53, 54, 55, 57
Volume VIII. NCRP Report No. 58
Volume IX. NCRP Reports Nos. 59, 60, 61, 62, 63
Volume X. NCRP Reports Nos. 64, 65, 66, 67
Volume XI. NCRP Reports Nos. 68, 69, 70, 71, 72
Volume XII. NCRP Reports Nos. 73, 74, 75, 76
Volume XIII. NCRP Reports Nos. 77, 78, 79, 80
Volume XIV. NCRP Reports Nos. 81, 82, 83, 84, 85

Volume XV. NCRP Reports Nos. 86, 87, 88, 89
Volume XVI. NCRP Reports Nos. 90, 91, 92, 93
Volume XVII. NCRP Reports Nos. 94, 95, 96, 97
Volume XVIII. NCRP Reports Nos. 98, 99, 100
Volume XIX. NCRP Reports Nos. 101, 102, 103, 104
Volume XX. NCRP Reports Nos. 105, 106, 107, 108
Volume XXI. NCRP Reports Nos. 109, 110, 111
Volume XXII. NCRP Reports Nos. 112, 113, 114
Volume XXIII. NCRP Reports Nos. 115, 116, 117, 118
Volume XXIV. NCRP Reports Nos. 119, 120, 121, 122
Volume XXV. NCRP Report No. 123I and 123II
Volume XXVI. NCRP Reports Nos. 124, 125, 126, 127
Volume XXVII. NCRP Reports Nos. 128, 129, 130
Volume XXVIII. NCRP Reports Nos. 131, 132, 133
Volume XXIX. NCRP Reports Nos. 134, 135, 136, 137
Volume XXX. NCRP Reports Nos. 138, 139
Volume XXXI. NCRP Report No. 140

(Titles of the individual reports contained in each volume are given previously.)

NCRP Commentaries

No.   Title

1   *Krypton-85 in the Atmosphere—With Specific Reference to the Public Health Significance of the Proposed Controlled Release at Three Mile Island* (1980)
4   *Guidelines for the Release of Waste Water from Nuclear Facilities with Special Reference to the Public Health Significance of the Proposed Release of Treated Waste Waters at Three Mile Island* (1987)
5   *Review of the Publication, Living Without Landfills* (1989)
6   *Radon Exposure of the U.S. Population—Status of the Problem* (1991)
7   *Misadministration of Radioactive Material in Medicine—Scientific Background* (1991)
8   *Uncertainty in NCRP Screening Models Relating to Atmospheric Transport, Deposition and Uptake by Humans* (1993)
9   *Considerations Regarding the Unintended Radiation Exposure of the Embryo, Fetus or Nursing Child* (1994)
10 *Advising the Public about Radiation Emergencies: A Document for Public Comment* (1994)
11 *Dose Limits for Individuals Who Receive Exposure from Radionuclide Therapy Patients* (1995)
12 *Radiation Exposure and High-Altitude Flight* (1995)
13 *An Introduction to Efficacy in Diagnostic Radiology and Nuclear Medicine (Justification of Medical Radiation Exposure)* (1995)

14 *A Guide for Uncertainty Analysis in Dose and Risk Assessments Related to Environmental Contamination* (1996)
15 *Evaluating the Reliability of Biokinetic and Dosimetric Models and Parameters Used to Assess Individual Doses for Risk Assessment Purposes* (1998)
16 *Screening of Humans for Security Purposes Using Ionizing Radiation Scanning Systems* (2003)
17 *Pulsed Fast Neutron Analysis System Used in Security Surveillance* (2003)
18 *Biological Effects of Modulated Radiofrequency Fields* (2003)

Proceedings of the Annual Meeting

No.   Title

1 *Perceptions of Risk*, Proceedings of the Fifteenth Annual Meeting held on March 14-15, 1979 (including Taylor Lecture No. 3) (1980)
3 *Critical Issues in Setting Radiation Dose Limits*, Proceedings of the Seventeenth Annual Meeting held on April 8-9, 1981 (including Taylor Lecture No. 5) (1982)
4 *Radiation Protection and New Medical Diagnostic Approaches,* Proceedings of the Eighteenth Annual Meeting held on April 6-7, 1982 (including Taylor Lecture No. 6) (1983)
5 *Environmental Radioactivity,* Proceedings of the Nineteenth Annual Meeting held on April 6-7, 1983 (including Taylor Lecture No. 7) (1983)
6 *Some Issues Important in Developing Basic Radiation Protection Recommendations*, Proceedings of the Twentieth Annual Meeting held on April 4-5, 1984 (including Taylor Lecture No. 8) (1985)
7 *Radioactive Waste*, Proceedings of the Twenty-first Annual Meeting held on April 3-4, 1985 (including Taylor Lecture No. 9)(1986)
8 *Nonionizing Electromagnetic Radiations and Ultrasound,* Proceedings of the Twenty-second Annual Meeting held on April 2-3, 1986 (including Taylor Lecture No. 10) (1988)
9 *New Dosimetry at Hiroshima and Nagasaki and Its Implications for Risk Estimates*, Proceedings of the Twenty-third Annual Meeting held on April 8-9, 1987 (including Taylor Lecture No. 11) (1988)
10 *Radon*, Proceedings of the Twenty-fourth Annual Meeting held on March 30-31, 1988 (including Taylor Lecture No. 12) (1989)
11 *Radiation Protection Today—The NCRP at Sixty Years*, Proceedings of the Twenty-fifth Annual Meeting held on April 5-6, 1989 (including Taylor Lecture No. 13) (1990)
12 *Health and Ecological Implications of Radioactively Contaminated Environments*, Proceedings of the Twenty-sixth

Annual Meeting held on April 4-5, 1990 (including Taylor Lecture No. 14) (1991)

13 *Genes, Cancer and Radiation Protection,* Proceedings of the Twenty-seventh Annual Meeting held on April 3-4, 1991 (including Taylor Lecture No. 15) (1992)

14 *Radiation Protection in Medicine,* Proceedings of the Twenty-eighth Annual Meeting held on April 1-2, 1992 (including Taylor Lecture No. 16) (1993)

15 *Radiation Science and Societal Decision Making,* Proceedings of the Twenty-ninth Annual Meeting held on April 7-8, 1993 (including Taylor Lecture No. 17) (1994)

16 *Extremely-Low-Frequency Electromagnetic Fields: Issues in Biological Effects and Public Health*, Proceedings of the Thirtieth Annual Meeting held on April 6-7, 1994 (not published).

17 *Environmental Dose Reconstruction and Risk Implications,* Proceedings of the Thirty-first Annual Meeting held on April 12-13, 1995 (including Taylor Lecture No. 19) (1996)

18 *Implications of New Data on Radiation Cancer Risk*, Proceedings of the Thirty-second Annual Meeting held on April 3-4, 1996 (including Taylor Lecture No. 20) (1997)

19 *The Effects of Pre- and Postconception Exposure to Radiation*, Proceedings of the Thirty-third Annual Meeting held on April 2-3, 1997, Teratology **59**, 181–317 (1999)

20 *Cosmic Radiation Exposure of Airline Crews, Passengers and Astronauts*, Proceedings of the Thirty-fourth Annual Meeting held on April 1-2, 1998, Health Phys. **79**, 466–613 (2000)

21 *Radiation Protection in Medicine: Contemporary Issues*, Proceedings of the Thirty-fifth Annual Meeting held on April 7-8, 1999 (including Taylor Lecture No. 23) (1999)

22 *Ionizing Radiation Science and Protection in the 21st Century*, Proceedings of the Thirty-sixth Annual Meeting held on April 5-6, 2000, Health Phys. **80**, 317–402 (2001)

23 *Fallout from Atmospheric Nuclear Tests—Impact on Science and Society*, Proceedings of the Thirty-seventh Annual Meeting held on April 4-5, 2001, Health Phys. **82**, 573–748 (2002)

24 *Where the New Biology Meets Epidemiology: Impact on Radiation Risk Estimates*, Proceedings of the Thirty-eighth Annual Meeting held on April 10-11, 2002, Health Phys. **85**, 1–108 (2003)

25 *Radiation Protection at the Beginning of the 21st Century–A Look Forward*, Proceedings of the Thirty-ninth Annual Meeting held on April 9–10, 2003, Health Phys. **87**, 237–319 (2004)

26 *Advances in Consequence Management for Radiological Terrorism Events*, Proceedings of the Fortieth Annual Meeting held on April 14–15, 2004, Health Phys. **89**, 415–588 (2005)

Lauriston S. Taylor Lectures

No.   Title

1 *The Squares of the Natural Numbers in Radiation Protection* by Herbert M. Parker (1977)

2 *Why be Quantitative about Radiation Risk Estimates?* by Sir Edward Pochin (1978)

3 *Radiation Protection—Concepts and Trade Offs* by Hymer L. Friedell (1979) [available also in *Perceptions of Risk*, see above]

4 *From "Quantity of Radiation" and "Dose" to "Exposure" and "Absorbed Dose"—An Historical Review* by Harold O. Wyckoff (1980)

5 *How Well Can We Assess Genetic Risk? Not Very* by James F. Crow (1981) [available also in *Critical Issues in Setting Radiation Dose Limits*, see above]

6 *Ethics, Trade-offs and Medical Radiation* by Eugene L. Saenger (1982) [available also in *Radiation Protection and New Medical Diagnostic Approaches*, see above]

7 *The Human Environment—Past, Present and Future* by Merril Eisenbud (1983) [available also in *Environmental Radioactivity*, see above]

8 *Limitation and Assessment in Radiation Protection* by Harald H. Rossi (1984) [available also in *Some Issues Important in Developing Basic Radiation Protection Recommendations*, see above]

9 *Truth (and Beauty) in Radiation Measurement* by John H. Harley (1985) [available also in *Radioactive Waste*, see above]

10 *Biological Effects of Non-ionizing Radiations: Cellular Properties and Interactions* by Herman P. Schwan (1987) [available also in *Nonionizing Electromagnetic Radiations and Ultrasound*, see above]

11 *How to be Quantitative about Radiation Risk Estimates* by Seymour Jablon (1988) [available also in *New Dosimetry at Hiroshima and Nagasaki and its Implications for Risk Estimates*, see above]

12 *How Safe is Safe Enough?* by Bo Lindell (1988) [available also in *Radon*, see above]

13 *Radiobiology and Radiation Protection: The Past Century and Prospects for the Future* by Arthur C. Upton (1989) [available also in *Radiation Protection Today*, see above]

14 *Radiation Protection and the Internal Emitter Saga* by J. Newell Stannard (1990) [available also in *Health and Ecological Implications of Radioactively Contaminated Environments*, see above]

15 *When is a Dose Not a Dose?* by Victor P. Bond (1992) [available also in *Genes, Cancer and Radiation Protection*, see above]

16 *Dose and Risk in Diagnostic Radiology: How Big? How Little?* by Edward W. Webster (1992) [available also in *Radiation Protection in Medicine*, see above]
17 *Science, Radiation Protection and the NCRP* by Warren K. Sinclair (1993) [available also in *Radiation Science and Societal Decision Making*, see above]
18 *Mice, Myths and Men* by R.J. Michael Fry (1995)
19 *Certainty and Uncertainty in Radiation Research* by Albrecht M. Kellerer. Health Phys. **69**, 446–453 (1995)
20 *70 Years of Radiation Genetics: Fruit Flies, Mice and Humans* by Seymour Abrahamson. Health Phys. **71**, 624–633 (1996)
21 *Radionuclides in the Body: Meeting the Challenge* by William J. Bair. Health Phys. **73**, 423–432 (1997)
22 *From Chimney Sweeps to Astronauts: Cancer Risks in the Work Place* by Eric J. Hall. Health Phys. **75**, 357–366 (1998)
23 *Back to Background: Natural Radiation and Radioactivity Exposed* by Naomi H. Harley. Health Phys. **79**, 121–128 (2000)
24 *Administered Radioactivity: Unde Venimus Quoque Imus* by S. James Adelstein. Health Phys. **80**, 317–324 (2001)
25 *Assuring the Safety of Medical Diagnostic Ultrasound* by Wesley L. Nyborg. Health Phys. **82**, 578–587 (2002)
26 *Developing Mechanistic Data for Incorporation into Cancer and Genetic Risk Assessments: Old Problems and New Approaches* by R. Julian Preston. Health Phys. **85**, 4–12 (2003)
27 *The Evolution of Radiation Protection–From Erythema to Genetic Risks to Risks of Cancer to ?* by Charles B. Meinhold, Health Phys. **87**, 240–248 (2004)
28 *Radiation Protection in the Aftermath of a Terrorist Attack Involving Exposure to Ionizing Radiation* by Abel J. Gonzalez, Health Phys. **89**, 418–446 (2005)

Symposium Proceedings

No. Title

1 *The Control of Exposure of the Public to Ionizing Radiation in the Event of Accident or Attack*, Proceedings of a Symposium held April 27-29, 1981 (1982)
2 *Radioactive and Mixed Waste—Risk as a Basis for Waste Classification,* Proceedings of a Symposium held November 9, 1994 (1995)
3 *Acceptability of Risk from Radiation—Application to Human Space Flight,* Proceedings of a Symposium held May 29, 1996 (1997)
4 *21st Century Biodosimetry: Quantifying the Past and Predicting the Future*, Proceedings of a Symposium held February 22, 2001, Radiat. Prot. Dosim. **97**(1), (2001)

5 *National Conference on Dose Reduction in CT, with an Emphasis on Pediatric Patients*, Summary of a Symposium held November 6-7, 2002, Am. J. Roentgenol. **181**(2), 321–339 (2003)

NCRP Statements

No.   Title

1 "Blood Counts, Statement of the National Committee on Radiation Protection," Radiology **63**, 428 (1954)
2 "Statements on Maximum Permissible Dose from Television Receivers and Maximum Permissible Dose to the Skin of the Whole Body," Am. J. Roentgenol., Radium Ther. and Nucl. Med. **84**, 152 (1960) and Radiology **75**, 122 (1960)
3 *X-Ray Protection Standards for Home Television Receivers, Interim Statement of the National Council on Radiation Protection and Measurements* (1968)
4 *Specification of Units of Natural Uranium and Natural Thorium, Statement of the National Council on Radiation Protection and Measurements* (1973)
5 *NCRP Statement on Dose Limit for Neutrons* (1980)
6 *Control of Air Emissions of Radionuclides* (1984)
7 *The Probability That a Particular Malignancy May Have Been Caused by a Specified Irradiation* (1992)
8 *The Application of ALARA for Occupational Exposures* (1999)
9 *Extension of the Skin Dose Limit for Hot Particles to Other External Sources of Skin Irradiation* (2001)
10 *Recent Applications of the NCRP Public Dose Limit Recommendation for Ionizing Radiation* (2004)

Other Documents

The following documents were published outside of the NCRP report, commentary and statement series:

Somatic Radiation Dose for the General Population, Report of the Ad Hoc Committee of the National Council on Radiation Protection and Measurements, 6 May 1959, Science **131** (3399), February 19, 482–486 (1960)

Dose Effect Modifying Factors in Radiation Protection, Report of Subcommittee M-4 (Relative Biological Effectiveness) of the National Council on Radiation Protection and Measurements, Report BNL 50073 (T-471) (1967) Brookhaven National Laboratory (National Technical Information Service, Springfield, Virginia)

Residential Radon Exposure and Lung Cancer Risk: Commentary on Cohen's County-Based Study, Health Phys. **87**(6), 656–658 (2004)

Index

Actuarial life tables 184
Age-specific death rates 8
Age-specific mortality risks 37
Aging, theories of 32
Argonne National Laboratory (ANL) 18
Atomic Bomb Casualty Commission 143
Atomic-bomb survivors 2, 4, 19, 48, 153, 172
Autocrine function 54

Bayesian 10–11, 16, 149, 177
Beagle(s) 18, 42, 80, 162, 167, 176
 animal of choice 80
 internally-deposited radionuclides 80
 osteosarcomas 80
Biodosimetry 13
Body burden ^{239}Pu 19
Bone, tumors or cancer, in 7, 77–83, 166–169, 175–181
 beagles 80
 external radiation 79
 humans 77–78
 induced by chemical carcinogens 79
 internally-deposited radionuclides 79
 mice 78–79
 oncogenes 79
 rats 79
 viruses 79
Breast cancer 6, 63–68, 153–156
 hormonal promotion 66
 ovariectomy 65
 sensitivity to carcinogens 64
 terminal end buds (TEB) 63
 totipotent stem cells and progeny 66
 Types 1–3 64
Bystander effects 104, 106, 109–110

Cancer pathology 43
Carcinogenesis 35
Carcinogenic process 39
 progression 39
 promotion 39
Chernobyl 164
Chromosomal aberrations 16, 42, 47–48, 50
 deletions 47–48
 Philadelphia (Ph1) 48, 50
Chromosome 4 42
Committee on the Biological Effects of Atomic Radiation 22
Conclusions 11–13
Cumulative survivorship curves 140–149

Deoxyribonucleic acid (DNA) 3, 12, 16, 24, 35
 damage 3, 12, 16
 microarray testing 24
 radiation-induced lesions 35
 repair 16, 35
Dog 18, 42, 80
Dose and dose-rate effectiveness factor (DDREF) 1, 13–14, 19
Dose, permissible 28
Dose-rate effectiveness factor (DREF) 1, 2, 4, 10, 12–13, 15, 19, 37, 149, 183
Dose-rate effects 16
Drosophila 26

Effects 1, 3, 25
  genetic 3
  somatic 3
  stochastic 1, 25
Effects of radiation 35
  increased when LET increased 35
  increase in dose increase 35
  reduced when dose rate reduced 35
Empirical methods 35
Erz Mountains 164
Exposures, radiation 15
  acute 15
  fractionated 15
  gamma rays 15
  homes 15
  mining population 15
  patients 15
  occupationally exposed 15
  x rays 15
Extracellular matrix 16
Extrapolation (history, radiation effects) 1, 8, 17–37
  across species, to humans 1, 17–37
  age-specific death rates 37
  insufficient data (present) 19
  life-shortening 37
  neoplastic 37
Extrapolation models 122–182
  absolute/relative risk 138
  acute 122–125
  age-specific mortality 138
  anticancer drugs 122
  body surface 123
  chemical toxicities 122–127
  chronic 125–127
  maximal tolerated dose (MTD) 123
  pharmacokinetics 123
  Phase 1 trials (animals) 124
  Phase 2 trials (humans) 123
Extrapolations, of radiation-induced 140–156
  breast cancer 149–156
  mice to dogs 140–143
  mice to dogs and humans 143–149
  mice to humans 149–156
  mortality 140–149

Gastrointestinal (GI) tract (cancer) 6, 75–77
  colon/colorectal 75–77
  esophagus 75
  rectum 75
  small intestine 75–76
  stomach 75–76
Gene 7, 16, 50
  alterations 7
  expression 16
  nomenclature (see Glossary)
  protooncogenes 50
  tumor specific 16
  tumor-suppressor 50
Genes, conservation of, for 35
  cell differentiation 35
  cell proliferation control 35
  DNA repair 35
Gene expression profiling 24
Genes (human) 24, 49–55, 72, 75, 77, 79, 94–96, 98, 101, 113–115
  ABL 51
  ADRA1 51
  ADRB2 51
  APC 77
  ATM 95, 98
  BCR/ABL (hybrid) 49, 51, 52, 55
  BRCA1 24, 94
  BRCA2 24, 94
  CD14 51
  CSF2 51
  FGF1 51
  GMF1 51
  HPRT 101
  IL3 51
  IL4 51
  IL5 51
  MYC 50
  NR3C1 51
  PBX 54
  RAS 113, 114
  RET 72

SPARC 51
TCF7 51
TLX1 54
TP53 50, 96
XPA 75
Genes (mice) 42, 52–53, 75, 77, 79, 94, 97, 114, 115, 116
　Abl 53
　Apc 77
　Bmi1 53
　Fos 79
　Hoxb8 53
　Hoxd3 53
　H*Ras* 114
　Il1b 53
　Il3 53
　Myc 52–53, 115–116
　Nbs1 94
　N*Ras* 53, 114–115
　Pax5 42
　Pbbcp2 53
　Pim1 53
　Pla2q2a 77
　Ptch+/− 75
　Ras 97, 114
　Tgfbr1 42
　Trp53 97
Genes (oncogenes) 49, 111–116
　Abl 49
　ABL1 116
　Myc 115–116
　Ras 111–115
Genes (tumor-suppressor) 117–120
　APC 118
　BRCA1 118
　BRCA2 118
　NF1 118
　NF2 118
　RB1 118
　TP53 117, 118, 119, 120
　Trp53 117
　Tsc2 119
　WT1 118
Gompertz 8, 27, 29–33, 130–138, 141, 145, 184
　distribution/function 29–33, 130–138, 141, 145
　equation 27
　intercept 32
　slope(s) 32, 33

Hematopoietic system (cancers) 5, 40–56
　atomic-bomb survivors 41
　cytogenetic processes 47
　Hodgkin's disease 41
　leukemias/lymphomas 40–42
　non-Hodgkin's lymphoma 41
　target cells 45–46
　trisomy, chromosome 21 41
Hematopoietic unit 44–45
　hematopoietics 44–45
　progenitor cells 44–45
　stroma 44–45
Hepatitis 39
　confounded by 39
Heritable diseases 1
　radiation-induced 1
Hogs 68
Homeostatic state 29
　departure from 29
Host factor(s) 5, 10, 36, 38, 40
　cell-cell communication 40
　cytokines 40
　differences 10, 36
　microenvironment 40
　stroma 40

Inhalation Toxicology Research Institute (ITRI) 168
Internal emitters 4, 11, 18–19, 78, 156, 160–164, 166, 168–169, 174–175, 177–178, 180–181
　Americium-241 164, 168
　Cerium-144 174
　Iodine-131 156
　Plutonium-238 168, 174–175, 178, 180–181
　Plutonium-239 11, 18–19, 156, 164, 168, 174–175, 177, 181
　Radon-222 156

Radium-224 19, 78, 156, 161–162, 166, 168–169, 177–178, 181
Radium-226 11, 18, 168–169
Radium-226,228 156, 160–162, 164, 168
Strontium-90 168, 174, 177–178
Strontium-90-Yttrium-90 181
Thorium-228 161, 164, 168
Thorium-230 163–164
Thorium-232 156, 161, 163
Yttrium-90 174
Yttrium-91 174
International Agency for Research on Cancer (IARC) 38
International Commission on Radiation Protection (ICRP) 19
Ions, heavy 2

Kaplan-Meier survival curves 142

Langerhan's cells 73
Lawrence Livermore National Laboratory 74
  workers 74
Lethality function 29
Life 8–9, 20, 27, 33–34
  mean duration of 27
  shortening 8–9, 20, 33–34
  span differences 34
Linear-energy transfer (LET) 2, 8, 12, 15, 17, 86, 106, 157, 159, 168
  high-LET 12, 15, 86, 106
  low-LET 8, 15, 17, 86, 106
Liver (cancer) 39
Lung (cancer) 4, 5, 56–63, 171–175, 181–182
  alveologenic 57
  adenocarcinoma 56
  atomic-bomb survivors 59–60
  bronchiolar 57
  genes *RAS* and *TP53* 59, 61
  hamsters 61
  human 56–62
  K-Ras 58
  large-cell carcinoma 56
  mice 57
  squamous-cell carcinoma 59
  radon induction 59
  rats 58
  Trp53 gene 58
  uranium miners 60

Manhattan Project 21
Mayak 168–169, 172
Mean after-survival (MAS) 33
Medical Research Council (MRC) 22
Messenger RNAs 24
Molecular/cellular radiation effects 84–91, 94–100, 102–104, 106, 109–120
  adaptive responses 110
  breakage first hypothesis 104
  bystander effects 109–110
  cell-cycle aspects 95–96
  chromosome aberrations 84, 102
  confounders 109–110
  double strand breaks 86–87
  DNA damage 85–86
  excision 91
  fluorescence *in situ* hybridization techniques 103
  genomic instability 110
  genes, DNA repair 94–100
  genes *XRCC2* and *Nbs1* 90
  high-LET 86, 106
  homologous recombination 88
  interchromosomal 90
  intrachromosomal 90
  low-LET 86, 106
  mutations 84
  non-homologous end joining (NHEJ) 88–89
  oncogenes 111–116
  recombination 90
  repair 86–91, 95–96, 109–116
  single-strand breaks 87
  tumor suppressor genes 117–120
Mortality 8, 26, 28, 32–33
  extrinsic 28
  intrinsic 8, 28
  laws of 26
  ratio 33
  theories about 32

Mutation(s) 16, 30–31
  life expectancy reduction 31
  rate(s) 31
  somatic 30

Neutrons 2, 12
  fission 2
Nontargeted radiation effects 106, 109–110
  adaptive 110
  bystander 106, 109–110
  genomic instability 110
Nuclear weapons 21
  radionuclide fallout 21
  testing 21

Oncogene activation 111–116
Oncogenesis 15
  human prediction 15
Opossum (*Monodelphis domestica*) 74

Patients 2, 15, 21
  radiotherapy 2
Plutonium 15, 21
  (see *internal emitters*)
Probability density function 10
Proportional hazard models (PHMs) 8, 11, 145, 150, 177, 181

Radiation 2, 4, 17
  external 4
  general population 17
  internal 4
  limits 17
  protection 2
  working population 17
Radiation Effects Research Foundation 143
Radiation weighting factors 1, 13, 15
Radionuclides 156
  (see *internal emitters*)
Radium dial painters 18, 160
Radon 11, 18 (see *lung cancer*)
  lung cancer 18
  uranium miners 18

Rat (strain) 64
  Sprague-Dawley 64
Recommendations 13–14
Relative biological effectiveness (RBE) 4, 12–13, 19, 37, 156
Risk 2, 10, 12, 21–26
  genetic 2, 21–26
  mortality 10
  relative 12
Risk estimates, of cancer induction 3, 22–37
  cell type 3
  direct method 23
  doubling-dose method 22–23
  gene number method 23
  history of somatic 26–37
  somatic 22–37
Rotifer (*Proales*) 26

Saint Bernard (dog) 80
Scaling factors 145
Sheep 68
Skin (epidermal) cancer 6, 73–76
  atomic-bomb survivors 74
  basal-cell carcinoma 6, 73
  guinea pigs 74
  humans 6, 73–75
  melanoma 6, 74
  mice 74
  opossum 74
  squamous-cell carcinoma 73
  tinea capitis patients 74
  ultraviolet radiation 75
  U.S. Department of Energy workers 74
Smokers 57
Stochastic effects 1, 157
Somatic genetic repair 7
  DNA 7
  DNA repair 7
  DNA replication 7
  DNA content 7
Stroma 44–45, 54
Strontium 11
  (see *internal emitters*)
Summary 183–187
Summary of findings 3–11

Survival times, mean 29
Survivorship curves 26
  Drosophila 26
  humans 26

Theories of 32
  aging 32
  mortality 32
Thorium 11
  (see *internal emitters*)
Thorotrast® 169
  (see *internal emitters*)
Thyroid cancer 6, 68–72
  adults 68
  cellular origin: clonal 71
  children 68
  goitrogenesis 71
  histogenesis 69
  humans 68–69, 71–72
  RET (protooncogene) 72
  RET/PTC1 72
  thyroid stimulating hormone (TSH) 70
Toxicity ratio, method/approach 4, 18–19, 168, 175–176
  plutonium-239/radium-224 19
  relative 18

Toxicology 3
  chemical 3
  radionuclide 3
Transuranics 11
  (see *internal emitters*)

United Nations Scientific Committee on the Effects of Atomic Radiation (UNSCEAR) 22
Ultraviolet (UV) exposures 6
Uranium 21
  (see *internal emitters*)
Uranium miners 18, 172
  lung cancer 18
  radon exposure 18
U.S. Atomic Energy Commission 18
U.S. Department of Energy 74
  workers 74
U.S. National Academy of Sciences/National Research Council (NAS/NRC) 22

Weibull distribution 150
World Health Organization (WHO) 60